Life in Lakes and R

A volume in the *New Naturalist* series, a survey
of British natural history published by Collins
and edited by John Gilmour, Sir Julian Huxley,
Margaret Davies, and Kenneth Mellanby. The
New Naturalist has been described by the
Listener as 'one of the outstanding feats of
publishing since the war', and by *The Times
Literary Supplement* as 'a series which has set a
new standard in natural history books'.
Founded in 1945, it now contains more than 50
volumes, of which the following have already
appeared in Fontana:

Life in Lakes and Rivers

T. T. Macan M.A., Ph.D.,
E. B. Worthington C.B.E., M.A., Ph.D.

COLLINS
The Fontana New Naturalist

First published in the New Naturalist Series 1951
Revised edition first published in Fontana 1972
Sixth Impression June 1980
Copyright © in the revised edition T. T. Macan and
E. B. Worthington 1972

Made and printed in Great Britain by
William Collins Sons & Co Ltd

Contents

Plates

Figures

Editors' Preface

Dr Macan and Dr Worthington, the collaborators in this scholarly and interesting book, have been colleagues for many years. Both are connected with the Fresh-water Biological Association's laboratory on Windermere in the English lakes.

The subject of life in British lakes and rivers occurred to the Editorial Board early in the planning of the New Naturalist series. No sooner had the idea come to us than we invited Dr E. B. Worthington to be the author of the book. Shortly after he had accepted the invitation, however, he left Windermere to take up an important appointment in East Africa. He suggested, when this happened, that the major task of the book should be taken over by his younger colleague, Dr Macan, and that the final work should be a collaboration of the two of them. The happy result of this suggestion is now before the reader, who will agree that Dr Macan and Dr Worthington have written an admirably lucid and vital book on a somewhat neglected subject.

One of the points made by teachers of ecology in the last twenty years to their students is that the animal and plant community can be readily studied in a pond or a lake. That this is so the authors of this book, which is entirely ecological in its outlook, make quite clear. Moreover, in so far as it can be made simple they make it so. Nevertheless the reader, when he has finished the book, will realize that the solutions of many problems of life in lakes and rivers (solutions which have been often arrived at by workers at Windermere) have only served to create more problems – problems wider and more fundamental than perhaps anybody ever suspected, problems that reach far into the very structure of biology.

First appointed in 1935, Dr Macan returned to Windermere in 1946 after five years as specialist entomologist in the Army. Dr Worthington, well known on account of his explorations of the African lakes, came to Wray Castle as Director in 1937, when that post was first created. Since 1946 he has been Scientific Secretary to the East Africa High Commission, in which position he has been surveying all East Africa in order to ascertain how and where the resources of science might be used to promote prosperity and well-being.

THE EDITORS

Authors' Preface

It is becoming increasingly difficult to write a book which is not out of date in a number of minor, and perhaps a few major, particulars, because advances in every field are continually appearing in print and no-one can hope to keep abreast of all of them. The present authors do not venture to hope that they have not erred, but they have been in what are undoubtedly the most favourable circumstances for writing a book of this kind. Members of the staff of a freshwater biological station, they have been surrounded by colleagues each an expert in one of the fields touched in the following pages. Moreover, these colleagues have been willing to read through chapters on their subjects and draw attention to errors and defects. Mr H. C. Gilson has read chapters 1, 2, and 4; Dr C. H. Mortimer chapters 2, 4, and 9; Dr J. W. G. Lund and Dr Hilda Canter have assisted with the botanical parts of several chapters; Dr W. E. Frost has read chapters 11 and 12; Mr E. D. Le Cren chapters 12 and 13; and Dr C. B. Taylor chapters 14 and 15. Captain C. Diver, C.B.E., Director of the Biological Service, and Mr F. T. K. Pentelow of the Ministry of Agriculture and Fisheries have criticized chapters 8 and 14 respectively. It is appropriate to acknowledge at this point that the editorial board has made many helpful suggestions.

No less important is a lay impression, since it is not for specialists and professional biologists that this book is primarily written. For this we are indebted to Mrs Zaida Macan, who has read through the whole typescript; and to Mr Maurice Illingworth, who has read Chapter 12.

We record all this kindness with gratitude and take the opportunity to thank those who have helped us.

Introduction

It is interesting to speculate on the contents of an Atlantic Charter drawn up by any species of animal other than man. We may start by comparing the lot of man with that of the rest of the animal kingdom, which is separated by a lower grade of intelligence. People are not often drowned as a result of catastrophic floods, few are blown to destruction by strong winds, and death in a forest or heath fire is a rare calamity; nor do abnormal spells of hot or cold weather claim many victims. The same is true for all large animals. But for innumerable small ones such dangers are great, and the populations of many are seriously depleted at intervals by one or other of these causes. A final calamity, which does not often befall man, though it befalls other animals of all sizes, is death at the hands of some beast of prey.

Medical science has rendered the more advanced sections of the community secure against many of the disease-causing parasites which formerly destroyed them in large numbers; plague (the black death), typhus (gaol-fever), and cholera may be mentioned as diseases that once took a heavy toll of life in the British Isles, but do so no longer. In less advanced parts of the world, disease still brings death to enormous numbers of human beings, and in this respect there is not so much difference between man and the other animals. Indeed on theoretical grounds it can be argued that many other animals are better off than man. Most parasites can survive only in living tissue and therefore it is to their advantage that the host should remain not only alive but unhandicapped in the struggle for existence. Too virulent a strain will be as unsuccessful as the one that is not aggressive enough to gain a foothold in the face of the counter-measures taken by the host. There is continual selection of a strain that can establish itself but will not kill. Selection acts on the host too and specimens lacking resistance to a particular parasite are removed at an early age. The result is a state of tolerance, with the host carrying parasites but not inconvenienced by them. This state is not attained by man for two reasons. First, selection of resistant individuals does not take place because medicine prevents it. Secondly, whereas the total population of most animals is separated into numerous units between which there is little interchange, man travels all over the world and carries strains of parasites from a region where tolerance has developed to one where it has not.

Death from starvation may have two causes: the food of a particular species may fail on account of some climatic irregularity, or may be eaten up by some other animal. We avoid the first of these eventualities by means of a highly organized system of transport; when the harvest fails in one part of the world, food is brought from somewhere else. But the recollection of the Bengal famine will serve as a reminder that this danger is not wholly eliminated. Against animals which eat the same food as himself, man brings to bear all the resources of science, and wages a never-ending war on rabbits, rats, grain-weevils, slugs, caterpillars, and a host of smaller pests. Lower animals cannot attack their competitors on anything like the same scale; until recently it might have been stated that they could not do it at all, but now it is know that certain organisms can produce chemical substances that kill competitors – penicillin produced by the mould *Penicillium* is the obvious example – and this process is comparable with the steps that man takes to safeguard his food supplies.

Were we writing about philosophy and not natural history, we should of course have to insert a passage about the perils peculiar to civilization – death on the roads from motor vehicles, and other accidents with machines, destruction by high explosives and other weapons of war, and mass annihilation by atomic bombs. However, we are not. The purpose of what has been written is to stress that the life of a small animal – and it is with such that this book is mainly concerned – is a continual struggle of extreme severity. The physical environment, predators, parasites, and competitors all have to be contended with. The response has been steady change and continual modification. Some animals have won and held a place where physical conditions are easiest – but dangers from predators, parasites, and competitors consequently greatest. Others have become adapted to conditions where physical or chemical conditions make life difficult – but where accordingly there will be fewer other organisms to harass them. Fresh water provides some of the most striking examples of the latter.

It might appear at first sight that the gulf between land animals and water animals is great, and that the easiest way into fresh water is from the sea. But, although an animal or plant may pass from marine to freshwater conditions with no alteration of structure, the change confronts it with considerable functional or physiological problems. The concentration of salts is generally much lower in fresh water than in the sea and moreover liable to considerable variation according to rainfall and other factors; from the biological point of view the constancy of the marine environment is one of its most notable features. Further, conditions vary widely from one freshwater locality to another. An animal proceeding up the Hampshire Avon, for example, would find, if it turned aside into one of the tributaries coming from the New Forest, an acid water, poor in dissolved salts, very different from that which would surround it if it followed the main river to a source at the foot of the chalk downs. These chemical conditions have proved a barrier which only a few marine animals have surmounted. Some of the snails, the bivalve molluscs, the freshwater shrimps, and the fishes are the most familiar.

Actually it appears from an examination of the groups to which present-day freshwater plants and animals belong, that it has been easier to invade fresh water from land than from the sea.

The animals of marine origin occupy in fresh water the same sorts of situation that they occupied in the sea, and they have not changed greatly. As a result of isolation they are now quite distinct from their nearest marine relatives, but they present no peculiar freshwater facies. Some of the land-animals, too, have effected the change to fresh water with little alteration; others with no more than some general adaptation such as the conversion of appendages from legs to paddles. But some of the animals from the land, having once established themselves in fresh water, have become considerably modified to live in one particular and difficult part of the underwater world such as a torrent; others have achieved remarkable physiological adaptations, such as the ability to live without oxygen. It is among these specialists that we find the peculiar and characteristic freshwater types.

The main problem confronting an animal taking to the water is how to obtain its oxygen. Often the difficulty is not great, because many land animals live in damp places and have a moist surface. If they are quite small, this surface is all they require for respiratory purposes, and accordingly it does not matter greatly whether air or water is the medium beyond the layer of surface moisture. The problem is not quite as simple as this, but we need go no further for the present.

Other animals, whose land ancestors were probably less dependent on humid conditions, spend their lives in the water, but have developed a variety of methods whereby they can utilize atmospheric oxygen. The familiar water-beetles swim to the surface with the aid of their hind legs (which, with the transfer to water, have been modified into efficient paddles), and take in a bubble of air between their backs and their wing-cases. Some snails come to the surface to fill a lung. Mosquito larvae – the well-known wrigglers of the domestic water-butt – feed at the surface with breathing-tubes penetrating through to the air. The rat-tailed maggot has a remarkable telescopic appendage, so that it can walk on the bottom and keep in contact with the air at the same time. Other animals take in oxygen as a gas but nevertheless live perpetually submerged. Several groups – none of them very familiar to the layman – have a close-set pile of unwettable hairs. Withdrawal of oxygen for use by the body from the gas entrapped in this pile causes a partial vacuum, which is filled by oxygen dissolved in the water. Others tap the gas-filled tubes inside plants by means of some part of the body modified for the purpose.

Fresh water has presented invaders with a variety of problems, and no animal has solved all of them; none is found in all the habitats and most are confined within a rather restricted range.

The surface offers a rich hunting-ground, for many land animals fall into the water and are held there helpless by the force of surface-tension. Two groups of insects have mastered the art of living on the surface film, and are able to prey on these unfortunates. They are, however, confined to small pieces of water on sheltered bays, presumably because the effort of keeping station against the wind in the middle of a big sheet of water is too great. The pond-skaters have developed long legs and proceed somewhat after the manner of long-oared skiffs. The other group, the whirligig beetles, have shortened and flattened their legs till they resemble more the paddles of a canoe; by means of which they move over the surface with great rapidity. The way of life of these surface-dwellers calls to mind that of the wreckers who once gained a livelihood round our coasts, though they are not able to take measures to lure their victims to destruction.

Quiet shallow conditions, where mud settles and rooted

plants – all incidentally of terrestrial origin – provide shelter, oxygen, and food, are probably the most easy to colonize; that is, they present the would-be inhabitant with least in the way of physical and chemical difficulties. But life is not easy in this habitat because there are so many different types of organism and competition between them is severe. The water-beetles and water-boatmen pursue their prey through the underwater jungle with the speed and grace of terrestrial felines; dragonfly nymphs lurk concealed like a crocodile in a waterhole; the tiny *Hydra* trails its tentacles in the water like a fisherman setting out his net. Against these marauders the caddis-larvae seek protection within a cumbersome house of stick or stone, and snails can withdraw into their more neatly made shell. But the snail's shell, the caddis-larva's house, the dragon-fly nymph's protective coloration, and the beetle's speed may alike prove unavailing when fish come nosing through the undergrowth in search of food. They are not the only enemies, and some birds and the water-shrew also hunt in this territory.

In deeper water, if it is too dark for plants to grow, the mud offers a substratum which is often rich, because the remains of animals and plants rain upon it from above. Here the chief inhabitants are mussels, worms, and midge-larvae, most of which are modified for burrowing and for feeding on minute particles. The diversity of form is not great, though the number of individuals may be colossal.

In yet deeper water conditions may be extremely difficult, because there is no oxygen for part of the year, but some animals, notably midge-larvae, have solved this physiological problem.

In the open water also the variety of form is not great, though numbers may be. Some small animals, such as the water-flea, resemble their counterparts in the sea, but the marine zoologist inspecting a freshwater catch is immediately struck by the lack of diversity. This is partly because many marine animals which lead a fixed or relatively sedentary life have a free-swimming young stage. This presumably serves the end of dispersing the species. In the circumscribed conditions of fresh water such a stage is of less advantage for this purpose, and might indeed prove a danger by being carried away to the sea before it was ready to settle. Freshwater animals which have come in direct from the sea have almost all lost the free-swimming stage, which their nearest marine relatives possess. Another striking feature of the animals of the open water is their small size. This is probably due to the absence of shelter and consequent vulnerability to predation. Some protection is obtained by transparency, a feature seen in both the sea and in fresh water, but the main defence is small size, rapid reproduction while conditions are favourable, and the formation of a resting stage as soon as they cease to be. Large animals could not reproduce fast enough to make good losses due to predation. It may be objected that some animals floating in the open sea attain a large size. In comparison with the sea most bodies of fresh water are very small and fishes feeding on the larger animals on the bottom in shallow water could easily make excursions to prey on any large organism that developed in the open water. Most of the sea is so far from land that this cannot happen, and predators and prey must develop in balance together.

One representative of that enterprising group the insects inhabits open water, and provides one of the most remarkable examples of adaptation that the animal kingdom has to show. It is the larva of a mosquito-like fly, and is known as the phantom larva, on account of the transparency already mentioned. The breathing-tubes, in other insects continuous from back to front, are vestigial except for paired air-sacs fore and aft. These are hydrostatic organs by means of which the animal can maintain itself at any desired depth. The earliest workers supposed that it functioned like a submarine, pumping fluid in or out of ballast tanks as required. Later observers, noting that the bladders never contained fluid, postulated that they worked like the swim-bladders of fishes, and secreted or absorbed oxygen. Finally, when a technique for analysing the minute quantities of gas in the bladders was evolved, it was discovered that the walls of the bladders expand or contract as the result of nervous stimulation, and the gases dissolved in the body-fluid diffuse in or out accordingly.

Rocky shores of lakes are startlingly barren places compared with those of the sea; there is no canopy of weeds nor incrustation of barnacles and other sessile animals. The two are not strictly comparable, because in fresh water there is no tide and the water may sink slowly to a low level and stay there during a long spell of fine calm weather. But even below this level there are not many living organisms. Probably this is due to a poor food-supply, for hard rocks and waters that are not rich in nutrient substances commonly go together. Perhaps this is an environmental niche that has not been completely filled; the

fact that neither rocks nor wood are attacked by boring organisms in fresh water, as they are in the sea, suggests that this idea
is not as revolutionary as it may seem at first. Further, the
zebra mussel, the only freshwater bivalve that can attach itself
to a hard substratum as the marine mussels do, has only recently entered fresh-water.

On a rock-face in a lake the only plants are green algae, and
the only animal the freshwater limpet. This is actually descended from terrestrial stock, though superficially it resembles
the marine limpet closely except in size. Smooth rock in a
stream may have some covering of moss, which harbours many
animals. If it is bare, it may be covered by great numbers of
buffalo-gnat larvae, which spin a web across its surface and
attach themselves by means of a circle of hooks on a basal pad.
They obtain food by straining the current with hairy mouth-
appendages.

Smooth rock is not found very frequently, and, where wave
action or running water prevents the settling of finer particles,
the bottom usually consists of stones and boulders. Several
animals have adapted themselves to these particular conditions. Some mayfly nymphs, such for example as those of the
March Brown, have flat bodies and strong claws, and can
crawl over the surface of a stone, where they graze on the
attached algae, in such a way that the current cannot get beneath them and pluck them off. Certain caddis-larvae spin nets
between the stones in a stream and subsist on the debris which
these nets strain from the current.

These animals are specialists. They have solved the main
problem of life in swiftly flowing water – anchorage; and two,
having surmounted this difficulty, have turned the peculiarity
of the medium to their own advantage – the constant flow
brings them their food. The modifications of the specialists
render them unable to compete with other animals except in
the habitat to which they are adapted. Except in extreme
conditions, specialists and non-specialists are found side by
side. In streams, for example, many animals without any particular modifications for life in running water occur beneath
the stones. One of them, the larva of the daddy-long-legs, is
not greatly different in structure from its relative the leather-
jacket, which lives in the soil and damages lawns and pastures
by eating the grass-roots. Another, one of the commonest, is
the freshwater shrimp (*Gammarus pulex*), a rather incom-
petent swimmer which is washed away at once if caught by the

current. It is one of the most successful of all freshwater animals. It may be abundant in quiet weed-beds, and sometimes, apparently when fish are very few or absent, it may live in the open water of lakes. But it is not ubiquitous and some of the chemical problems posed by fresh water have proved too much for it; it is not found where the calcium concentration is very low, and it also requires a relatively large amount of oxygen in the water.

There is another difficulty with which fresh water confronts animals and plants that seek to live in it – it may dry up. Many freshwater organisms have, accordingly, developed a resting stage, which is resistant to desiccation and probably plays an important part in dispersal as well as survival. Some animals have specialized in the temporary habit, thereby gaining certain advantages, for they make a quick start when water reappears, and can exploit the resources of the pond before it fills up with competitors coming from permanent pieces of water. Many mosquitoes lay drought-resisting eggs in damp hollows, and no development takes place till the hollows fill with water. That beautiful animal, the fairy shrimp, is found only in temporary pools, and survives the dry periods in the egg-stage. In England it is rare, but in Iraq, for example, where there are innumerable pools that fill when the high river-level causes the water-table to rise in spring, but are dry throughout the rainless summer, it is one of the commonest pond animals.

Nobody in these islands has far to go to find a piece of fresh water, and his search will usually take him into pleasant surroundings. He will not find the gaily coloured, almost gaudy, creatures that the seaside naturalist encounters; most freshwater animals are rather drab. On the other hand neither land nor sea animals can so easily be kept and watched at home. Moreover a day's pond-hunting is not rendered fruitless by rain, as is an excursion after butterflies and many other land creatures. The fauna is, as we have just seen, the product of a difficult environment, a severe struggle for existence, and the adaptability and plasticity of living organisms. In consequence it shows a fascinating diversity of form and function, which cannot fail to appeal to all who are interested in wild creatures and how they live.

1. First Principles

Life in lakes and rivers is studied by three different sorts of naturalist, whose spheres of interest all too rarely overlap. First there is the naturalist with a pond-net who collects the smaller animals and plants of the water; secondly there is the naturalist with a fishing-rod, who classes organisms according to their relationship to fish; finally there is the naturalist with field-glasses for whom rivers, lakes, and reservoirs are places where interesting birds and mammals may be observed. This book attempts to link these three fields together, to relate them to the geographical background, and to discuss the conflict which is bound to centre round them in a thickly populated and heavily industrialized country.

The fisherman is often less interested in the question of what kinds of animals and plants occur in a given set of conditions than in the question of how many, or rather how much, and this question is also fundamental for naturalists of the other two classes. The answer, as given by the study of productivity, has provided a central connecting theme in the pages that follow. Productivity must ultimately depend on the amount of non-living material brought into a body of water in solution and in suspension, and so the story starts with the geology of the surrounding land. Closely bound up with this is the way in which the body of water was formed. When the primary nutrient materials reach the water they are utilized by plants, and plants always form the first link of the biological food-chain. Succeeding links are provided by various invertebrate animals and finally fish. At all stages living organisms die and decompose, resolving eventually into the simple substances from which they are built up. The mechanism of this process is studied by the bacteriologist. The result presents the chemist with many of his problems, and takes him particularly to the mud over which the water lies. This may be at a considerable depth and far below the point to which light penetrates sufficiently to make plants grow. The return of these substances to the upper layers, obviously of great importance to the biological cycle, involves an incursion into the field of physics.

Fish with predaceous habits are generally the final link of

the foodchain within the water itself. But the chain continues on to the land and into the air, for there are piscivorous birds and mammals which the freshwater biologist must study.

Man himself exerts a profound effect on life in fresh water. First, since earliest times he has striven to keep water within defined limits by means of drains and flood banks. More recently he has taken to using rivers as convenient agents for the removal of his waste products. Sewage, in moderate amounts, enriches water as it enriches land, but in excess it uses up all the oxygen in solution, with disastrous effects on most living organisms. In conflict with those who have wastes to dispose of are those who wish to see their waters well stocked with fish. These people, too, have played a big part in altering the conditions of water-life. Finally, man is today faced with an ever-increasing problem of obtaining for domestic use water containing the minimum possible amount of life.

Many insects start life in the water but end it as terrestrial creatures with the power of flight. Their life-histories have fascinated naturalists for a very long time and they attracted the attention of some of the first workers in the field of freshwater biology. But studies involving all the animals, all the plants, the chemical and physical background, and the inter-relationships between them were not made until later. To work of this kind the terms limnology, hydrobiology, or freshwater biology are applied indifferently. The pioneer was Professor F.-A. Forel who, in 1872, settled down to a lifetime's investigation of Lake Geneva or Le Léman. His main publication runs to three volumes, the first of which appeared in 1892, and considers the lake from thirteen different aspects.

In 1884 scientific investigation of fishery problems with a view to legislation started in Hungary, and in the two succeeding decades research stations with the same object sprang up all over Europe. In 1890 Professor Otto Zacharias started a station for fundamental research at Plön in Schleswig-Holstein. It was a private venture but it was supported by the State. With its foundation Germany took the lead in both applied and theoretical research and she retained this position until the recent war. Professor August Thienemann was director at Plön during the period between the wars, and his name is associated with many limnological studies, particularly those relating to the classification of lakes according to oxygen concentration, and according to the species of midge (Chironomidae) found in the bottom mud. Stations for fundamental research were started in other countries in the years which followed, usually in connection with universities. The best known today are: Hillerød in Denmark, opened in 1900 by the University of Copenhagen and made famous by Professor C. Wesenberg-Lund, who devoted the first ten years to plankton problems and has since studied the biology of many invertebrate groups; Lunz, opened in 1905 on an Alpine lake in Austria; Aneboda, started in 1908 by the University of Lund in Sweden and associated particularly with the name of Einar Naumann, who elaborated theories of lake classification; and Tihany on Lake Balaton in Hungary. The Istituto Italiano di Idrobiologia on Lake Maggiore is a later foundation, but no less celebrated than the other stations mentioned, particularly for the study of plankton. Limnology was also studied at universities and at stations devoted to the practical study of the production of fish.

Development in America was similar. The first station was founded in 1894 by the University of Illinois, but the most famous of the early contributions to theoretical studies were made by C. Juday and E. A. Birge at the University of Wisconsin.

Since 1945 expansion has been rapid all over the world, and it is impossible to give any general account of it, one reason being that, whereas before the war few scientific communications were published except in one of the major languages of western Europe, now they appear in a great many. Development in Britain has probably been similar to that in many other countries, and we may therefore pass on to events there.

Great Britain lagged far behind in the early years and it is interesting to quote the words of Professor Charles A. Kofoid (1910) who, in 1908 and 1909, toured the research stations of Europe. He writes: 'The direct support of biological stations by educational funds of local or state origin, often in connection with universities, so generally prevalent in other European countries, is almost wholly lacking in Great Britain.'

'The stations have been forced, therefore, to turn to memberships of supporting societies composed to a considerable extent of scientific men themselves, to private benefactors and to the commercial interests of the fisheries for aid. The result has been a relatively meager and fluctuating financial support ... and a relatively very large absorption of the funds and activities of the British stations in scientific fisheries work.'

However, in spite of this, or should it be because of this, the 'meager and fluctuating financial support' having deterred all but the most determined and enthusiastic from seeking employment of this sort, Kofoid's opinion is: 'The scientific fisheries work done by the British stations is unsurpassed in its excellence and effectiveness.'

Marine problems have always taken pride of place in Britain. As befitted the leading maritime power of the day, she was the first to send out a major expedition to explore the depths of the ocean, when it was first realized that life existed there. H.M.S. *Challenger* set out early in 1873 and was at sea until late in 1876. Soundings, collections of animals and plants, chemical analyses, meteorological records, and other scientific data were obtained in all parts of the world, and the total achievement was considerable. It was the culmination of a collaboration between science and the Royal Navy which had been yielding fruit for a century or more. One of the junior scientists was a certain John Murray, who later, as Sir John Murray, became head of the Challenger Office, and was responsible for seeing the final volumes of the reports through the press, many years after H.M.S. *Challenger* had been relegated to the scrapyard.

He found time to organize Britain's first important contribution to limnology – the Survey of the Scottish Freshwater Lochs, carried out between 1897 and 1909. This was a private venture undertaken after he had ascertained that the Lords Commissioners of the Admiralty were not concerned with fresh water, and that the Survey Department of the Office of Works (late Ordnance Survey) was not interested in anything except the surface of bodies of fresh water. Five hundred and sixty-two lochs were surveyed and, though sounding was the principal activity, sufficient observations on temperature were taken to provide the data for a theory about the circulation of water in the deeper parts of lakes – a theory which is still accepted today. Plankton collections were made but other biological observations were few.

In 1901 Mr Eustace Gurney started a station on Sutton Broad in Norfolk, and during the succeeding years a vigorous programme was carried out here under the direction of his brother, Dr Robert Gurney. It was, however, a private venture and it lapsed when the Gurney brothers moved away from the neighbourhood.

The next event of importance in the history of British

freshwater biology was the issue in 1915 of the final report of the Royal Commission on Sewage Disposal, appointed in 1898. This Commission had carried out a careful examination of almost all aspects of the problem, even to the extent of inaugurating research to obtain certain information which it deemed essential. That our rivers are still polluted by sewage must be laid at the door of the legislators and not blamed on the Royal Commission.

Between the two wars several rivers were surveyed by staff of the Ministry of Agriculture and Fisheries. The primary object was to discover the effect of pollution on animals and plants, but obviously in order to do this it was necessary to survey unpolluted stretches for comparison, and the result was an important contribution to knowledge about the fauna and flora of uncontaminated rivers. During the same time Dr Kathleen Carpenter of the University College of Wales investigated stream faunas and the effect on them of pollution from lead mines. This work established a tradition of freshwater biology in Wales which has persisted ever since.

In the twenties the foundation of a station for freshwater biological research in Britain was discussed. A number of distinguished men of science came together and worked hard exploring ways and means. The interest of universities, academic societies, fishermen, and waterworks undertakings was aroused and, when, in 1930, subscriptions totalled £575 and promise of a grant of about the same amount had been obtained from the Government, the time to start was deemed to have arrived. Ideas about a new, properly equipped, building had to be abandoned, and search was made for an existing building which could be adapted. It had been decided that Windermere was the most suitable location, and on the banks of this lake the committee found that a place called Wray Castle was only partly occupied. It appeared to be suitable and in October, 1931, work started in a Victorian country house built externally in the style of a medieval castle.

At the beginning there were two naturalists, P. Ullyott and R. S. A. Beauchamp, and one laboratory assistant, George Thompson. The apparatus and general facilities were meagre, as may be appreciated from the amount of money available. Further subscriptions were raised, and by the end of 1935 there were five research workers and three assistants. In the following year a committee from the Development Commission inspected the laboratories and the work in progress, and, as a result of their visit, a bigger annual grant from the Treasury became available. One of their recommendations was the appointment of a full-time director, this office having previously been honorary and filled by Dr W. H. Pearsall (later Prof. W. H. Pearsall, F.R.S.), at that time Reader in Botany at Leeds University.

Expansion continued and in 1947 there were ten research workers, twelve laboratory assistants, and an instrument-maker. Wray Castle was now too small to provide, as it had done hitherto, laboratories, and living accommodation for unmarried members of the staff and visiting research workers, and in 1948 what had been the Ferry Hotel was purchased to take its place. The move was effected in 1950 and now, twenty years later, the staff of nineteen scientific officers and fifty-two supporting staff is once again complaining of lack of space. It is planned to build an annexe. Most of the staff came from Cambridge in the early days, and at that time there were few other universities from which they could have come. If W. H. Pearsall was the father of limnological thinking in Britain, a cofounder of the Freshwater Biological Association, J. T. Saunders, was the father of limnological teaching. Among the students who attended the course he ran, in addition to those mentioned, were F. T. K. Pentelow, B. A. Southgate and C. F. Hickling, three pioneers whose names will be encountered later in this account, and G. E. Hutchinson, who is likely to be the last person to write a comprehensive treatise on limnology single-handed.

Students come to the laboratory of the Freshwater Biological Association every Easter to attend a course. For reasons of accommodation and transport, numbers are limited to about sixteen. During the first few years applications were often fewer than this. After the war freshwater biology became increasingly popular at universities, and demands for places on the course rose, which frequently meant that from five students specializing in freshwater studies two were selected. This situation proved unacceptable to teachers who, one by one, organized their own courses. Today (1970) there are universities where the numbers on the course are well above sixteen. It is the universities and colleges where freshwater biology is not taught that now send most of the students to the Freshwater Biological Association's course. Cambridge is one of them.

During the decades since the war university expansion has

provided posts for freshwater biologists, and some of the leading men have, unfortunately, found that American universities offer better conditions than those in Britain. A station for research on fish was established at Pitlochry by the Scottish Home Office soon after the war, and in 1962 work started at the Freshwater Biological Association's River Laboratory in the south of England. Increasing numbers of freshwater biologists have also found employment with River Authorities, with which statement we conclude this brief review, for we lack faith in our ability to forecast the future.

As a final introductory topic, physical and chemical properties which affect living organisms demand brief notice. Warm water is lighter than cold water and so floats on it, a phenomenon which leads to temperature-layering of lakes in summer, and thereby exerts a profound effect on the animals and plants. This subject is explored further in later chapters.

In the present chapter we shall notice only some of the properties of water at low temperatures. Water is densest at four degrees above freezing point on the Centigrade scale. This is 4° C., since freezing point is at 0° on this sensible scale, but 39.2° on the Fahrenheit scale, which is still in common use in Britain, and on which 32° is the freezing point of pure water.

As the surface of a lake cools down in the autumn, the upper layers sink and displace warmer water from below. This process goes on till the temperature is uniform at 4° C. from top to bottom. Water colder than 4° C. is less dense and therefore floats at the surface, and, if there is no wind to stir it up and mix it with the water below, this surface layer will be quite thin. Further cooling leads to the formation of ice. There can then be no physical mixing due to wind and, if cold conditions at the surface persist, the effect can only pass through the water by the slow process of conduction. In Britain, therefore, ice never gets very thick.

If water were to become steadily denser until freezing point was reached, a body of water would attain a condition where the temperature was uniformly just above freezing point from top to bottom. Further cooling would presumably cause the whole mass to freeze solid. It has been stated in print that such a state of affairs would mean that nothing could live in fresh waters in temperate latitudes. This is hardly likely to be true because a number of animals can withstand being frozen solid, but it is certainly more convenient, particularly for man, that water is heaviest at 4° C.

For every thirty feet that an object sinks below the surface of the water the pressure upon it increases by one atmosphere. The pressure in deep water has been brought vividly to the notice of many a biologist who has inadvertently lowered a water sampler unopened into deep water, and hauled it up to find quite flat what had been a cylinder. Water itself is almost incompressible, and, if it were quite incompressible, Windermere, which is 219 feet or 67 metres deep at the deepest point, would be only a millimetre or about 1/25th of an inch deeper than it is at present. There is not, therefore, a big increase in density with increasing depth and no grounds whatever for the popular idea that objects thrown overboard in deep water do not go right down to the bottom, but float at a certain depth, light objects reaching a point of equilibrium before heavier objects; anything of higher specific gravity than water will go on sinking till it reaches the bottom. The pressure inside an aquatic organism is approximately the same as the pressure outside, and creatures which live in deep water do not, therefore, possess adaptations to withstand pressure as is sometimes supposed. Rapid progress from deep to shallow water may prove disastrous for any animal, because bubbles of gas appear in the blood on account of the reduced pressure; swim-bladders of fish may burst.

Water is twice as viscous near freezing point as at ordinary summer temperature, and this has an important bearing on the rate at which small bodies sink.

The surface tension of water is a physical factor which looms very large in the lives of animals and plants below a certain size. Some animals such as the water-crickets can support themselves on the surface of the water by it, and snails and flatworms can sometimes be seen crawling along the underside of the surface film. Occasionally aquatic creatures get trapped in the surface film and are unable to get back into the water. Terrestrial animals that alight on the water surface frequently find themselves entrapped, and at certain times of year these unfortunates make quite an important contribution to the food supply of certain predators which dwell in or on the water.

Any natural body of water will contain a certain amount of dissolved matter, the quality and quantity of which will depend on the geology of the land over which or through which the water has flowed. It is possible to recognize certain types, though generalizations are not very profitable because modifying factors are numerous. The main substances in solution in some of the chief types of water are shown in Table 1.

Table 1. The metallic and acidic radicles of the commoner dissolved substances in certain natural waters: figures in parts per million.

	Ennerdale, Cumberland	Cambridge tap-water	Braintree, Essex	Burton Well	Sea Water	The Dead Sea
Sodium	5·8	1·8	343	51	10,720	14,294
Potassium		7·5		57	380	4,418
Magnesium	0·72	26·5	15	39	1,320	41,324
Calcium	0·8	51	17·5	159	420	17,153
Chloride	5·7	18	407	90	19,320	181,960
Sulphate	4·6	16	96	378	2,700	623
Carbonate	1·2	123	218	121	70	trace
Silicon dioxide	1·6	—	2	9	1·4	trace

Ennerdale is an extreme example of a soft water. Cambridge tapwater is a fairly typical hard water derived from a drainage area in which there are chalk downs. The radicles present in much greater amount in the Cambridge water than in the Ennerdale water are calcium, magnesium, and carbonate. The Burton well-water is included as a curiosity which may be of interest to beer drinkers; it has an unusually large number of radicles present in high or relatively high concentration. The permanent hardness of Burton water is due to gypsum – calcium sulphate. A chloride content higher than usual is commonly due to spray from the sea, to wind-blown sea-sand, or to pollution. In inland areas well away from any maritime influence the chloride content is often examined as a routine part of the test for pollution.

If a water containing calcium carbonate flows through a soil containing sodium, sodium displaces calcium and the calcium goes out of solution. An example of such a water is that from Braintree in Essex shown in column three of Table 1; water draining from a calcareous region passes through the Thanet Sands, which are marine in origin, and emerges with quite a small amount of calcium in solution. This displacement of calcium by sodium is the essence of the 'Permutit' process for water-softening. Incidentally hard waters are frequently softened before being supplied to consumers. This is now a practice at Cambridge and its tap-water today contains less calcium than is shown in Table 1, in which the figures are from an analysis made before the softener was installed. Very soft waters, on the other hand, are sometimes treated with lime in the belief that defective teeth in the local children are due to the low calcium content of the water; but no convincing proof that this is so has ever been given. Very soft waters sometimes corrode pipes, owing to the presence of humic acids, and this can be cured by adding lime.

Further figures may be found in Taylor (1958), where there are seventy pages of them, not only from all parts of the British Isles but from other parts of the world as well.

The sea contains the accumulation of salts brought down by fresh waters over a period of aeons. Calcium has been lost from sea-water generally not by precipitation but by incorporation into the skeletons of animals, which have later died and fallen to the bottom of the sea. Small, single-celled animals play a greater part in this process than larger ones; for example, *Globigerina* ooze, which covers vast areas of the bottom of the ocean, is made up chiefly of the calcareous shells of a small single-celled animal bearing that name. Present-day chalk downs were formed under the sea by the accumulation in this way of the skeletons of myriads of tiny animals.

A similar concentration of salts takes place in lakes occupying areas of inland drainage, where there is no outlet and the water lost by evaporation is equal to the amount flowing in. In some such lakes the process has gone further than in the sea. Common salt or sodium chloride is the most abundant chemical substance in the sea; but the Dead Sea has reached a stage where there is some precipitation of sodium chloride, and this substance is present in smaller amount than the more soluble magnesium chloride. The proportions of these two salts in the River Jordan are the reverse of those in the Dead Sea. But there are no drainage areas in Britain without egress to the sea, and therefore discussion of such places is outside our present scope.

There are many substances present in water in very small quantity. It is known that on land and in the sea some of these so-called trace elements are important biologically and the same is probably true in fresh water.

No mention has been made so far of nitrates and phosphates, which are usually present in fresh water. As will be seen in a later chapter (see Fig. 2, p. 39) they are essential for plant growth, and during the course of it their concentration in the water is reduced. The fluctuation throughout the year is large and a single value for any one piece of water is, therefore, of no great significance.

Finally, of extreme importance to living organisms is the amount of dissolved gases in the water. Under average conditions at 0° C. (32° F.) there will be about 10 cubic centimetres of oxygen and half a cubic centimetre of carbon dioxide dissolved in one litre of water, that is 100 parts and 5 parts per million respectively. The concentration falls with rising temperature and at 20° C. (68° F.) there will be only about 65 parts of oxygen and rather less than 3 parts of carbon dioxide. For certain purposes it is convenient to express the concentration as the percentage of the saturation concentration at the temperature prevailing when the sample was taken. Fifty parts per million of oxygen would be 50% of the saturation value at 0° C. but 77% at 20° C.

Animals use up oxygen and produce carbon dioxide and plants do the same in the dark. While illuminated, the latter do the reverse, absorbing carbon dioxide and producing oxygen. Still water with much vegetation in bright sunlight may for a period have more oxygen in solution than the normal maximum at the temperature prevailing. This condition, which is unstable, is technically known as super-saturation.

Decomposition also uses up oxygen, and serious pollution, by sewage for example, exerts its effect on the fauna by depleting the water of oxygen.

One rather important point is that, if water is quite still, oxygen or any other substance in solution can only pass from a region of higher to a region of lower concentration by diffusion, and this process is extremely slow.

2. A Typical Lake

Warm water floats on cold water. If two layers differ markedly in temperature, the difference in density is such that even considerable disturbance will not mix them. However the opening sentence is true only down to 4° C. At lower temperatures cold water floats on warm water. The result of this peculiar property of water is that the lakes of the temperate region, with which we are concerned here, become stratified in winter and in summer. Because of the greater difference in summer it is the stratification during this period that is the more important biologically.

The left-hand side of Figure 1 shows the actual state of affairs in Windermere in February 1948; the temperature is uniform at 4° C. from top to bottom. Incidentally, a fact not always appreciated by dinghy sailors is that these cold conditions may persist well into March. By this time the sun is rising higher each day and shining for longer, and it starts to warm the upper layers – only the upper metre or two because the heating part of its rays is soon absorbed in water. The first fine spell is probably followed by windy weather, and the warm water at the surface is mixed with the colder water below; the lake is once more at a uniform though slightly higher temperature. Sooner or later, however, the two layers are established

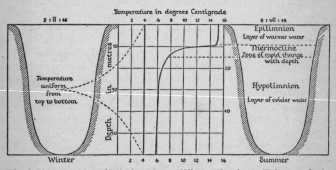

Fig. 1 Temperature of Windermere at different depths on February 2nd, 1948, and July 8th, 1948 (from data supplied by Dr. C. H. Mortimer)

with such a big temperature difference between them that they remain separate for the rest of the summer.

Once it is firmly demarcated, the warm upper layer of water increases in temperature relatively rapidly, and it also increases in depth because, when disturbed by the wind, it mixes with eddies of cold water from below. The cold lower layer has no source of heat, except perhaps from a small amount of mixing with warm surface water. The right-hand side of Figure 1 shows that by mid-summer the upper layer is many times warmer than in winter, but the lower layer has increased in temperature by no more than two degrees. Between the two there is a short depth of water in which the temperature drops rapidly. The warm upper layer is known as the epilimnion, the cold lower layer as the hypolimnion, and the region of rapidly dropping temperature between them is the thermocline. Greek scholars will have no difficulty with these terms, others may be puzzled to remember which of the first two is which, but there is a simple mnemonic, for *epi-* and *upper* begin with vowels and *hypo-* and *lower* with consonants.

Wind is the next factor. Blowing over the surface of a body of water it will set up a current which carries water to the leeward side. Obviously there must be a compensatory return current. This will flow along the bottom of the epilimnion where it floats on the hypolimnion, and therefore the wind keeps the epilimnion in constant circulation. There will be some eddying and turbulence and this will keep the water of the epilimnion thoroughly mixed. In Britain totally windless periods seldom last long.

Chemical analysis shows that the water of the hypolimnion is well mixed, for the concentration of dissolved substances is the same at all depths. Dr C. H. Mortimer, F.R.S., made a thorough investigation of this phenomenon in Windermere and eventually provided an explanation. He then devised a model which demonstrated the explanation in a very convincing way. It represented the longitudinal section of a lake enclosed between two sheets of plate glass. Two wires ran the length of the section near the surface, and an electric current passed through them warmed the adjacent water to create an epilimnion. A dye was carefully run into this to mark it. A gale was created by two blowers originally designed to dry ladies' hair. Water is so heavy that a real gale tilts the water surface so little that only the most sensitive apparatus records a rise in level at the leeward end, but the thermocline tilts considerably. In the model the epilimnion was blown towards the leeward end and the dyed water was displaced into the form of a wedge. If the wind was strong enough the thin end of the wedge did not reach the windward end; in other words the hypolimnion was exposed at the surface here and temperature measurements have shown that this happens in a real lake. Some of this surface hypolimnion water is mixed with the epilimnion which, when the wind ceases, is accordingly deeper and colder. However, the way in which the dyed water in the model retained its identity on top of the clear water was striking. When the wind drops the epilimnion flows back until the thermocline is level once more, but its momentum causes it to overshoot, the epilimnion piles up at the other end and the thermocline is tilted in the opposite sense. This seiche, as it is called, continues for days in a real lake, the angle at which the thermocline tilts decreasing with each oscillation. In the north basin of Windermere an oscillation takes about nineteen hours. On the right-hand side of Figure 1 the epilimnion is of uniform temperature down to a depth of nearly ten metres and this is the condition found after thorough mixing by strong wind. In calm weather the temperature tends to decrease, often in an irregular manner, from the surface to the thermocline. As the hypolimnion flows to and fro with each oscillation, irregularities on the bed of the lake set up eddies, and these produce the mixing of the water which led to the investigation originally.

The rivers and streams flowing into a lake are usually at a temperature well above that of the hypolimnion and accordingly will mix only with the epilimnion.

With the shortening of the days in autumn, particularly if there is a fine spell with cold clear nights, heat is lost by radiation during the hours of darkness. The epilimnion begins to cool down and eventually, sometimes not until December, a gale will obliterate it, mixing it completely with the hypolimnion.

There remains one other factor to mention before anything living comes into the picture. The sun's rays have been considered so far only from the point of view of their heating properties; for the activities of plants light is more important. Light rays do not penetrate far into even the purest water, and in most waters there is something extraneous to reduce their penetration still more. Any sedimentary matter in suspension, any colouring such as that derived from peat, and living organisms themselves all absorb rays of light. Figure 11 (p. 72) shows that in three Lake District lakes light goes farthest into the pure and barren waters of Ennerdale Lake, less far into the richer waters of Windermere, and least far into the peat-stained and rich waters of Bassenthwaite. Since light does not penetrate far into water, plant growth is only possible in the upper layers, and is nearly always confined to the epilimnion.

May we recapitulate here, since so much of what follows depends on the physical conditions which have just been discussed. During the summer months the lake is divided into an upper warm epilimnion and a lower cold hypolimnion, which are to all intents and purposes completely separate (Fig. 1), and all plant growth takes place in the epilimnion.

Algae (minute floating plants) are present in the open water all through the winter but physical factors, notably the short days and the low light intensity, are unfavourable for rapid multiplication. When conditions are right for this there is a rapid and colossal increase in numbers which is checked when the substance in shortest supply is exhausted. Phosphate and

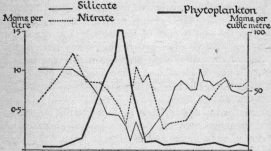

Fig. 2 Increase in phytoplankton and decrease in the concentration of certain salts in Windermere in 1936

nitrate are two important nutrients but in Windermere, the size of the population of *Asterionella*, the commonest diatom, is limited by the concentration of silica, which the alga requires for its skeleton. Once reproduction is halted, the population declines rapidly (Fig. 2). After the spring outburst various species of algae rise and fall in numbers, but the total attained is much less than that reached in the spring. The zooplankton (small floating animals) reach their maximum abundance a month or two later than the phytoplankton.

The animals living in the mud at the bottom of the lake are in perpetual darkness and almost constant temperature. Little is known of their activities in any British lake, but P. M. Jónasson has shown that in the Danish Esrom lake the growth of a chironomid depends on the rain of dead plankton falling from above. This comes to an end in winter and the growth of the larvae stops. It starts again in the spring and proceeds rapidly, but is checked again when the oxygen is used up in the lower layers and the larvae can do little more than survive. They emerge early in the following year. Most larvae take two years to complete development but a few achieve it in one, but their eggs are all eaten by their brothers and sisters who have failed to develop as fast. The result is a big emergence every other year. Nearly all aquatic insects emerge as adults in spring or summer, presumably because of the physiological difficulties of flying in cold weather, and this must impose a seasonal rhythm upon their development.

A ring of green algal growth on the stones in the shallow water of a lake appears in spring, but most of the stoneflies and some of the Ephemeroptera of this region grow during the winter and pass the warm part of the year in the egg stage. This phenomenon will be discussed further when streams are described. One of the commonest animals in the reed-beds is *Leptophlebia* (Ephemeroptera) and this is a species that grows throughout the winter, but most of the fauna grows during the summer.

These various plants and animals are continually dying and decomposing, broken down by various agencies about which we do not know very much at the present time. Fungi and bacteria set upon their dead bodies and reduce them to fine particles and simple compounds, which serve as food for other organisms, so that there is a constant process of breaking down and building up in the epilimnion. But some of the decaying fragments, with the organisms breaking them down, fall through to the hypolimnion, and we must leave them for the moment to describe what has been happening there. More important perhaps is what has not been happening; there has been no plant growth, because it has been too dark, and therefore no utilization of the dissolved substances for want of which algae have been dying in the layers above. Evidently division into epilimnion and hypolimnion reduces the productivity of a lake.

The decaying matter which falls down to the hypolimnion continues to decay, though at a slower rate on account of the low temperature, and it uses up oxygen. There is no source from which the oxygen in the hypolimnion may be replenished, and consequently the concentration falls steadily all through the summer; it may reach nil if the lake is a productive one and the hypolimnion small – an important point, as will be seen in the next chapter.

The decaying matter may eventually reach the bottom, and here some of it is eaten by the animal inhabitants of the bottom mud, and some of it is broken down into simple substances by bacteria and other agents. Most of the organic matter found deep in the mud, where it must have lain for thousands of years, was washed in from the land. But these simple substances cannot reach the surface layers, where they could be used for building up more living matter, until hypolimnion and epilimnion mix in the autumn. By then biological activity is reduced, and by the time there is a big demand again for dissolved nutrients in the following spring, much of the supply will have been washed out of the lake. On the average, water takes nine months to pass through Windermere, and therefore during the winter there will be considerable depletion of the dissolved substances released from the hypolimnion by the autumn mixing. Again it becomes apparent that the formation of a hypolimnion prevents the development of the full potentialities of a lake.

Large fragments hardly decay at all in the cold mud at the bottom of deep lakes. Wasmund (1935) gives an account, illustrated by gruesome photographs of bodies, including three human ones, that have been brought up, generally in fishermen's nets, after many years in the water.

Dr C. H. Mortimer (1941–42) has recently shown that, when there is oxygen at the surface of the mud, iron is present in the oxidized ferric state and forms a colloidal complex with various other substances. This colloidal complex tends to hold the simple products of decay, and therefore augments the locking-up process caused by the slow decomposition in the mud. But, if all the oxygen is used up, the ferric iron is reduced to the bivalent ferrous state, which goes into solution with consequent breakdown of the colloid complex. This liberates the other substances, and Mortimer was able to show, both in an artificial experiment in an aquarium and in a lake, that the disappearance of oxygen from the hypolimnion is followed by an increase in the concentration of silicate, phosphate, ammonia, and iron in the water.

The above are factors which affect the plants and animals living out in the open water of a lake and in the mud below it.

A different assemblage of living things inhabits the shallow regions near the shore, and this population too is affected by physical and chemical processes. The most important is wave-action. The effect of this factor depends on the nature of the land on which it acts. Waves beating upon rock will disintegrate the weaker patches and leave the harder ones projecting as ridges but the total effect is small; waves beating upon sand or peat, on the other hand, will erode the shoreline rapidly. Many lakes are surrounded, partly or entirely, by moraine deposits known by various names such as glacial drift, boulder clay, till, or sammel. Waves eroding a shore of this type leave *in situ* only the larger stones and boulders and carry away the finer particles, which eventually come to rest in deeper water away from the shore, or in some sheltered bay. The coarsest particles will be moved the least, the finest the greatest distance, and there will therefore be a graded series passing into deeper water farther away from the shore. The processes of erosion and deposition result in what is known as a wave-cut platform and are illustrated diagrammatically in Figure 3.

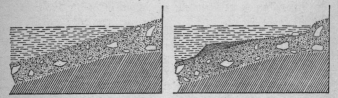

Fig. 3 Diagram of the erosion of a boulder clay shore to give a wave-cut platform

Sometimes the material removed is not carried out at right angles to the shore but at an acute angle so that, when it settles, it forms a spit. Such formations are of importance to animals and plants because they create areas of quiet water which are the resort of certain species unable to tolerate the conditions on a wave-beaten shore.

Deltas are even more important features of the lake shore.

They may be no more than bulges in the shoreline, or, at the other extreme, they may cut a lake in two. Good examples of deltas at all stages are to be seen in the Lake District lakes. The delta of the Measand Beck stretched two-thirds of the way across Haweswater, before this lake was dammed in 1941 to provide more water for Manchester. A stage farther can be seen in the valley of Buttermere and Crummock Water, which were left by receding glaciers as one large lake. Since then Sail Beck, flowing in from the east, has cut the original lake into two and its delta now provides the half-mile of flat land in the valley floor between the two lakes. Another pair of lakes, Derwent Water and Bassenthwaite, show a still more advanced stage. Here again the two were formerly one, but the River Greta has poured so much silt and gravel into the original lake that there is now a full two and a half miles of plain separating the north shore of Derwent Water from the south of Bassenthwaite.

Some of these deltas are much too large to have been brought down by the little streams existing today, and much of the material was probably swept down during the last stages of the Ice Age by the bursting of ice dams and other minor cataclysms.

We may pass from generalities to describe a portion of the shoreline of Windermere, for it illustrates several of the points already made, and is referred to later when the fauna is discussed. The shoreline in question is bounded to the north by a ridge of rock jutting into the lake. The sides of this promontory, which is known as Watbarrow point (Fig. 4) are smooth, and run down at a steep angle to a depth of nearly 100 feet. To the south the same kind of rock, Bannisdale slate, is exposed at the edge of the lake, but weathering and wave-action have broken it up considerably, and the products of its disintegration litter the lake floor. They are large flat angular slate-like stones. Moon (1934), who has studied this region of the lake, refers to it as the 'Bannisdale' shore and contrasts it with the 'drift' shore which lies to the south. The drift shore consists of stones and boulders but these are round, not flat and angular, and there are finer particles between them. This shore has been formed by the erosion of a mound of boulder clay or glacial drift, somewhat after the manner shown in Figure 3. The hinterland of the Bannisdale shore is covered by woodland, but that of the drift shore has been cleared of woodland at some time and is now pasture. This is not coincidence;

where the underlying slate is not covered with glacial drift, the topsoil is often so thin, and rocky outcrops are so frequent, that cultivation of the land is not feasible; but where the rock is covered by boulder clay, it has been worthwhile to remove the forest and bring the land into agricultural use.

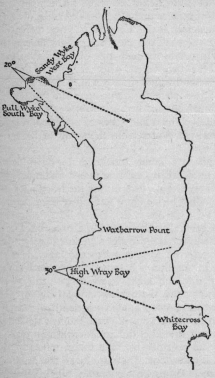

Fig. 4 Windermere, north end showing reed-beds. Reed-beds are stippled

Figure 4 shows that at the south end of the drift shore there is a bay – High Wray Bay – which is somewhat protected. Only the comparatively rare easterly gales will blow right into it, and the range of direction of wind from which it gains no protection at all is but 30°. High Wray Bay is floored with sand.

Sandy Wyke Bay farther north is more sheltered. The range of direction of wind which will blow straight into it is only

20°. But a glance will show that the amount of exposure is not to be measured entirely by the angles drawn in Figure 4. If a wind blowing in the direction of the more southerly of the two pecked lines bounding the High Wray Bay angle veer slightly, it will still drive waves into part of the bay, and it must shift through nearly another 30° before complete protection is obtained. But if a south-easterly wind that just blows full into Sandy Wyke veer but a few degrees, the projecting coastline will shelter the bay almost completely. Sandy Wyke Bay is also sandy, but there is a big reed-bed growing in it.

Only a north wind will blow right into Pull Wyke South Bay, but it will traverse so short a stretch of water that the waves raised will not be of significant size. This bay is floored with fine mud. The vegetation shows the zonation typical of quiet conditions. In the shallowest water there are various emergent plants such as reeds, rushes, sedges, and horsetail; in deeper water there are plants with leaves floating at the surface, such as water lilies; and beyond them are plants, such as pondweeds, stoneworts, and quillwort, which live totally submerged throughout life.

The phenomena described above are of such general occurence that, in spite of the diversity of lakes, a 'typical' lake is a useful concept. There are two main types of lake that may be styled 'atypical'. Lakes that have a large surface area and little depth do not stratify. Lough Neagh, possibly even Bassenthwaite and Derwent Water in the Lake District, are examples. Lakes of this type, however, have not been studied thoroughly. and all that can be said at present is that they have been found to be unstratified in summer at a time when epilimnion and hypolimnion are clearly demarcated in other lakes. Of course, any body of water in temperate climates will show some stratification after a hot day; the important point is how long stratification lasts. It is possible that these large shallow lakes may stratify throughout an occasional summer when sunshine is unusually abundant and wind unusually scarce. Information should be available from Lough Neagh soon, as the New University of Ulster has established a station there. It is difficult to make observations sufficiently often unless a laboratory is available, and the ideal, described in the next chapter, is an arrangement of thermometers in the lake connected to a recorder in the laboratory.

The other kind of atypical lake is known technically as meromictic, and its peculiar feature is permanent stratification.

The density difference that prevents mixing is due to substances in solution, not to temperature, and is often but not invariably due to peculiar geological conditions. The condition could arise in any lake where production is high and circulation low. Poor circulation occurs in areas where strong wind is rare, and the effect of lack of wind will be enhanced in a lake with a small surface area relative to its depth, and with not much water flowing in. An abrupt transition from winter to summer and from summer to winter is another factor that plays a part. Given these conditions one may postulate that the meromictic condition arose in the following way. If at the end of a summer the hypolimnion is greatly enriched by decomposition of organisms produced in the upper layers, it will be denser than the water from which they have come when both layers are at the same temperature. It is not difficult to suppose a year in which the cycle of events has resulted in both being at 4° C. The epilimnion will float on the hypolimnion. If there is little wind and ice forms soon, this state will endure until the spring. If there is little wind then to upset this delicate state of balance, and plenty of sun to increase the density difference by warming the upper layers, stratification will have lasted a year. By the following autumn the accumulation of two years' production will have increased the density difference due to solutes between hypolimnion and epilimnion and the chances of their remaining unmixed during the following season are greater. The longer the two remain separate the more the energy required to mix them, and the less likely mixing becomes. It is believed by Professor I. Findenegg, who discovered the condition, that certain lakes in the Carinthian province of Austria became meromictic in some such way.

So far no definition of the word 'lake' itself has been attempted. Our colleague, Mr F. J. H. Mackereth, has been heard to say that a lake is no more than a bulge in a river. This idea is more useful to a chemist than to a biologist, but it is salutary that a biologist should remember how much of what takes place in a lake is governed by what is washed in from the drainage area. A lake is a piece of water of a certain size but at what size the word pond becomes applicable is a matter of opinion. It is one of those continuous series, frequently encountered in biology, where the difference between two ends is enormous but any lines drawn in between them to separate categories are arbitrary. One definition maintains that any piece of water which is so shallow that attached plants can grow all over it is a pond. The pedant has no difficulty in picking holes in this definition and pointing out that the depth to which attached plants extend varies very much with the transparency of the water; a cattle pond only a foot deep may be without vegetation in the middle because the light is cut off by innumerable small organisms which live in the open water and batten on the nutrients supplied by the dung. Or the nature of the substratum may be unsuitable for attached vegetation. Another school holds that, if a body of water becomes divided into epilimnion and hypolimnion and remains so divided throughout the summer, it is a lake and not a pond. Stratification, however, depends, not on size, but on the relation of depth to surface area and also to exposure to wind. The latter also determines to some extent whether the edges are eroded by wave action or not, and therefore blurs the definition according to which a lake is large enough for its shores to be eroded and a pond is not. In a restricted area, or for a given purpose, a worker may find a useful distinction between a lake and a pond, but in general no scientific distinction can be made.

3. Apparatus for Studying Lakes

Anyone provided with a stout net, some bottles, and a white dish or sheet can do an immense amount of work in fresh water. He can wade as far as is necessary into many ponds and streams and collect in the shallow water of lakes. He can even collect the plankton from the open water of a lake, if a suitable point of vantage is to be found. However, more serious work on the open water and any kind of work on the fauna of the mud or of the submerged weeds requires more elaborate apparatus. The first necessity is a boat. If work is to be done in deep water a winch is desirable. A very useful type of light winch which can be put on to any row-boat is made by the firm of Friedinger of Lucerne. It has the advantage that the wire is paid out over a pulley block of special circumference to which is attached a cyclometer, so that the depth at which the instrument hangs below the surface is shown accurately to the operator in the boat. With such a winch the different instruments for measuring temperature or light intensity, and for collecting water samples or plankton, can be lowered easily to any depth. It is often necessary to operate an instrument at a considerable depth in the water before hauling it back to the surface, and this may be achieved by despatching a so-called messenger down the wire. The messenger is usually a lump of metal with a hole drilled through it; on reaching the instrument at the bottom of the wire, it strikes some projection which is arranged to release a catch in order to perform the necessary operation.

An example is provided by the reversing thermometer. This thermometer is mounted on a pivot about its middle, and the pivot has a spring which turns the thermometer upside down when the catch at the top of the frame is released by the messenger. This reversal breaks the mercury column, and so, when the thermometer comes to the top, it shows the temperature at the depth at which the messenger struck it; warmer water through which it may pass leaves it unaffected. The ordinary clinical thermometer works on somewhat the same principle.

The thermometer has to be specially built to resist the high pressure which obtains under water and so it is a comparatively large instrument, which will not immediately take up the temperature of the surrounding water. Accordingly it has to be left for a few minutes at each depth from which a reading is desired, and, since, further, it must be hauled to the surface to be read, the taking of a series of observations is a long process. It is still used on expeditions and long excursions, but for regular work it has been obsolete for some years. The popular device at present contains a substance whose electrical resistance changes considerably with a relatively small change of temperature. It requires, therefore, a battery and a galvanometer but, when these are available and transport presents no problems, the apparatus, known as a thermistor, is convenient. Much of Dr Mortimer's work, described in the preceding chapter, was carried out from a boat, but latterly he had a series of thermistors slung at intervals between the bottom and a buoy moored in the deepest part of the lake. Each was connected to a recorder in the Ferry House, and what amounted to a continuous record was obtained. Dr Mortimer had nothing to do except convert the readings to °C. and work out what was happening. One of the authors was once explaining to a group of visitors what the recorder indicated and had just got to the point where emphasis is laid on the fact that the bottom of the lake is always cold when, by unfortunate coincidence, Mortimer, out on the lake, started to haul his line of thermistors up to the surface.

Apparatus of a somewhat similar kind is used for measuring the amount of light penetrating below the surface, which, we have seen, is so important in determining the depth of plant activity. A photoelectric cell, contained in a pressure casing, is connected to the surface by wires, and a window facing upwards is inserted into the pressure casing so that rays of light penetrating from the surface can strike the cell. They cause a small electric current, varying in amount according to the intensity of the light, and this can be measured in much the same way as with the temperature apparatus by a galvanometer in the boat.

When measuring sub-aqueous light, it is necessary to lower the instrument from a long support projecting sideways from the boat, because otherwise the boat would shade the instrument hanging beneath it. Not only the general intensity of light below the surface, but also the kind of light, is of great importance. This can be determined with the same instrument by covering the window with filters of various colours.

There is a much simpler but useful instrument for giving a rough idea of the clarity of water, known as Secchi's disc after the scientist who first used it. This consists of a white plate of 20 cm. (8 in.) diameter, which is lowered below the surface to the point at which it becomes invisible to the naked eye. This is, of course, a crude way of measuring how far light can penetrate, but Secchi's disc is very easy to carry about and use, and is accurate enough to provide comparisons between different types of water.

For most kinds of chemical work on water, and also for studying microscopic life, it is necessary to obtain samples of water from different depths. Here again the simple expedient is adopted of despatching a messenger down the wire to close a water-bottle at the desired depth. A variety of different kinds of water-samplers are used for this purpose. A simple example is a metal cylinder open at both ends so that when it is lowered it will pass through a column of water without disturbing it much. It is halted at the required depth and a messenger is sent down the wire. This releases lids which close over the top and bottom of the cylinder and are kept tightly in place by strong springs. The apparatus is now watertight, and can be hauled to the surface with a sample of water from the depth at which it was closed.

This self-closing metal water-bottle is an excellent instrument for many purposes, but for the study of bacteria, of which very many kinds inhabit fresh water, it is no use. The spores of bacteria are everywhere – in the air, in the water, on one's fingers – and accordingly a water-sampler for bacteriological investigations has to be arranged so that every part of the instrument which comes in contact with the actual sample of water collected can be sterilized by heat and kept in a sterile condition until the sample enters it. The principle was therefore adopted of using glass sampling vessels of a simple and standard pattern, held in a metal framework fitted with the necessary gadgets to operate an opening and closing device. The bottle is sealed with a bung pierced by two tubes. One is long and runs down to the bottom of the flask, and the other ends flush with the inside of the bung and is bent into an S-shape outside. A U-shaped piece of glass rod fits into a length of rubber tubing attached to each tube. When the bottle is at the required depth, a messenger is sent down to release a strong spring which pulls the glass rod out of the two tubes. Water runs down into the flask through the long tube, driving air out of the other tube until the flask is completely filled. A bubble of air remains in the bent tube so that no mixture can take place between the sample in the flask and the surrounding water during haulage of the whole apparatus to the surface. In practice, a number of flasks, each with its stopper and tubes, are sterilized in the laboratory and then a series of samples for bacteriological examination can be taken at different depths or at different places during the same outing.

Water may be obtained from any depth by lowering a tube and sucking. In the early days of the Freshwater Biological Association, when lack of money placed a premium on ingenuity, Mortimer used a bicycle pump with the washer reversed to obtain samples. If two bottles are connected in series, with the larger nearer the pump, the smaller and the tube will have been sufficiently washed by the time the larger is full. Water can also be raised by a stream of air bubbles emitted from a small tube inside, and extending almost to the lower end of, a larger one.

Plankton is commonly caught by means of a conical net made of material woven in such a way that the holes retain their size. A mesh of 60 meshes to the inch is generally used for animals, one of 180 meshes to the inch for algae. The efficiency of a net falls as the catch blocks the pores and for quantitative work the amount of water that has passed through the mouth must be measured by means of a propeller attached to a recorder. Many methods of catching plankton have been tried, particularly at the station at Pallanza, and the quest continues. One difficulty is that some of the animals swim away from an object they see coming through the water, or away from the pull of a current caused by suction into a pipe.

The easiest medium to sample is the mud on the bottom of a lake, though each sample must be subjected to a tedious process of sieving before the animals can be isolated. Often a simple tube will secure enough animals. If they are scarce, a larger sample may be obtained with a Birge-Ekman grab, which is a metal box open at the bottom and provided with two hinged lids at the top. Two jaws to close the bottom are held along the sides against the pull of strong springs. Going down, the apparatus passes through the water with little disturbance. This is important, for if there is obstruction the apparatus will not pass through the water, but push it aside, and it will also push aside the top layers of the mud if these are fine and fluid.

The lids fall when the box sinks into the mud and comes to rest. A messenger trips the bridle that holds the jaws up and the springs then pull them together to close the bottom of the box.

Stones and vegetation are less easy to sample quantitatively. Several workers have found that the number of animals caught in a given time or in a given number of sweeps of a net indicates, sometimes with unexpected accuracy, relative numbers in different places. Numbers per unit area can be calculated if samples with a quantitative sampler are taken in the same part of the lake at the same time. The Danish workers have used a square box open top and bottom to sample stony substrata near lake margins. It is placed over the bottom, stones are removed, and the water inside is baled out and poured through a net. This method cannot be used in running water because the box deflects the current downwards and causes it to scour the area that is to be sampled.

In Windermere H. P. Moon used a square frame on which he could pile stones to represent an area of natural substratum before lowering it onto the bed of the lake and leaving it until it had been colonized. The frame is one-third or one-half of a square metre in area and underneath it is covered with fine gauze to prevent the loss of animals while the frame is being hauled up. Stout wire-netting beneath the gauze adds additional support for the stones.

The Surber sampler is used by some workers to sample the stony substratum of streams and rivers. It consists of two frames, generally about one tenth of a square metre in area. These fold into one plane for transport and open at right angles for use. The horizontal one is placed on the bottom and the vertical one supports a net. Stones are then removed from the bottom inside the horizontal frame, and brushed in the mouth of the net to dislodge animals clinging to them, after which the remaining small stones, gravel and debris are stirred with a stick until it is believed that all living material has been swept into the net. We have not found this a satisfactory instrument because the current is often so swift that when one stone is picked up the stones above it shift to fill the gap. If the current is slow many good swimmers probably swim out of the net, if they are ever carried into it. More satisfactory, though not by any means free of error, is a shovel of some kind which can be pushed into the substratum for a known distance. Designs have ranged from a shovel with high sides with a net at the back, to a cutting edge connected to the handle by two strips which also support the frame of the net. A strong coarse net arrests the stones and a long tapering fine one any animals that have let go. If the stones are tipped into a solution of high specific gravity, calcium chloride or magnesium sulphate are suitable, the animals float to the top.

Weeds in rivers trail downstream and may be severed with shears and caught in a large bag. Another method is to hold a box with sharp edges a known distance above a lid and then bring the two together enclosing and severing the weed in a known volume. Weeds in still water rise vertically, and a device that cuts each leaf or stem as it meets it is preferable to one that pushes them downwards and does not cut them until they are pressed against the bottom. One such instrument consists of two tubes, about 8 cm. across, fitting one within the other. A boss on the inner passes through a slit in the outer and holds it in position, allowing a small amount of rotation to and fro. As the tubes are lowered into a weed-bed, the outer tube is rotated and the vegetation is severed between the sharp teeth which have been cut in the lower end of both tubes. They pass across each other like the teeth of a haycutter.

Incidentally parallel samples with this instrument and a net have shown the latter to be unexpectedly selective. It collects an unduly high proportion of species that tend to flee and an unduly low proportion of those which, like leeches, tend to cling to the substratum.

Larvae of Chironomids and many Trichoptera cannot yet be named, and in order to find out what species are present it is necessary to trap the emerging adults. In still water a box open at the bottom may be floated in a frame. The top should be of some transparent plastic material to keep the rain out, but at least one side should be of gauze to prevent condensation. Dr. J. H. Mundie has devised various modifications for use in both still and running water. In a lake he used conical traps into the top of which a screw-top jar could be screwed. Entrance to it is through a cone which prevents the animals falling back into the water. The whole apparatus can be submerged, an advantage in a lake to which the general public has access. For use in streams he built a heavy trap that could be anchored to the bottom. Triangular in both plan and elevation it offered minimum resistance to the current, which tended to press it downwards. Three legs kept it raised off the bottom and the catch entered a screw-top jar as in the other model.

Much decomposition takes place in the top few centimetres of the mud, and substances diffuse from it into the water. A study of these processes, important to the general economy of the lake, requires a sample disturbed as little as possible. The Birge-Ekman grab does not bring up such a sample and for this purpose the Jenkin surface-mud-sampler was invented. It consists of a large glass tube about 6 cm. in diameter, held by a band about its middle on to a metal frame, standing on four spreading legs. The sampler sinks into the soft mud when lowered to the bottom, but without disturbing it, and then, a messenger sent down the wire having released a catch, two pairs of arms travel forward to place a cap on either end of the glass tube. The speed at which these caps move into position has to be very slow in order not to disturb the mud and water in the tube, and this is effected by means of a pressure chamber of the same kind as that used for preventing doors from slamming. When closed, the glass tube contains a sample of the top few inches of deposit together with the water above, and the caps at either end are held tightly in place by springs. At this point the whole apparatus is hauled to the surface gently to avoid disturbance. The glass tube is detached from the frame, and the sample of bottom deposit, complete with the water immediately above it, just as it was at the bottom of the lake, can be carried into a laboratory for chemical and other tests. The person who devised this most useful apparatus was a retired engineer, Mr B. M. Jenkin, and it is worthy of note that the first experimental model, which he made largely out of meccano, operated so well that it was still in frequent use at the laboratories of the Freshwater Biological Association at Windermere ten years later.

Mr Jenkin was set another and much more complicated problem, namely to devise an instrument capable of extracting cores from the bottom deposits in lakes, if possible to a depth of twenty or thirty feet below the mud surface. A good deal of trial showed that an ordinary open tube or pipe was useless for this purpose because it compresses and disturbs the layers of deposit too much. After some thought, Mr Jenkin hit upon the idea of a sampler which could be thrust into the deposit first and then made to carve out a core by means of a curved cutting blade working on a long pivot. The business end of the instrument, which cuts out the core, is about four feet long, and consists of a tube cut in half lengthways and covered with a metal plate except for a slit down one side. A second half-tube lies within the first, attached on an axis in such a way that it can be rotated out through the slit. When the apparatus has been driven to the required depth, the inner half-tube is rotated, its sharp leading edge passes out of the slit, and, travelling through 180°, comes up against the far side of the plate. Between the inner half-tube and the plate there is now a sample of mud isolated from its surroundings with the minimum of disturbance. The rotation of the inner half-tube is effected by a system of cogs and a driving-wheel worked by a wire from the surface. The cutter can be attached to a series of tubes so that it can be driven to the desired depth in the mud before it is operated. The force required to press the whole instrument into the bottom is provided by a series of heavy lead weights, of which the number is adjusted according to the depth at which the particular sample is required. Thus a complete core, say twenty feet in length, is obtained in a series of overlapping cores each four feet in length.

The method of using this instrument is briefly as follows. First a pontoon with a derrick is firmly anchored over the spot from which the core is to be taken. Next a flat weight on a thin wire is lowered to the bottom to serve as a guide and as an exact measure of the depth to which the main instrument is subsequently lowered. Then the coring machine itself is lowered from the derrick on a stout wire with a pair of arms clutching the aforementioned guiding wire. The machine is allowed to sink into the deposit to the required depth, when a sample is required from near the surface; for a deep core the machine is allowed to sink as far as it will go and driven the rest of the way. Then a messenger is despatched down the guiding wire in order to release the arms, and the guiding wire is hauled up, an action which also operates the machinery for cutting out the core. It remains for the whole machine to be hauled to the surface and laid flat before the half revolution of the cutting blade is reversed and the core is exposed ready for transfer to the laboratory. It will be appreciated that the successful handling of this apparatus is no mean task; in fact it requires a team of three or four operators well trained in the particular functions which each has to perform at the right moment. With its aid, however, a large number of cores, some of them covering twenty-one vertical feet of deposit, have been collected from many parts of Windermere, and these have provided valuable information about the history of lakes since the Ice Age.

Mr Jenkins' apparatus proved excellent for use on Windermere, where the necessary pontoons could be borrowed, but not elsewhere, and accordingly Mr F. J. H. Mackereth devised a portable model (Fig. 5). The problem was to ensure stability while the core was being obtained. This he solved by basing the corer on a large cylinder resembling a dustbin (G). This sinks some way into the mud when the apparatus has been lowered, and is then forced farther in by means of a pump (P) which removes the water from its upper portion. A secure base has now been secured for the rest of the operation. The core is obtained in a long tube (B) housed inside a second tube (A), which is attached to the centre of the top of the anchoring cylinder. The second problem was how to drive the corer into the mud from a small boat that could not easily be kept exactly above the apparatus. Compressed air passing down a flexible tube (O) was the solution. It involved a piston fitting inside the outer tube and closing the top of the inner one (C). Some means of evacuating the inner tube as it moved downwards was essential, for otherwise a solid cylinder rather than a tube was being forced into the mud. This is achieved by a fine central tube (D) which holds in position a piston (F) at the mouth of the inner tube when this is retracted, passes through the upper piston and out through the top of the outer tube (L) to which it is attached. When compressed air admitted to the top of the outer tube forces the inner tube down into the mud, the air in the inner tube escapes through the fine central tube and the corer passes into the mud, causing no more compression than is due to the friction of the walls. When the inner tube is nearly fully extended, the compressed air escapes into a side tube (I), which leads it into the anchoring cylinder. This is forced out of the mud and brings the whole apparatus to the surface. Compressed air passed into the fine inner tube brings the inner tube back into the outer and, at the same time, ejects the core. This apparatus has been carried to tarns in the mountains by a helicopter and successfully used there.

4. Different Kinds of Lakes

Lakes, geologically speaking, are transitory features of the landscape. The biologist who studies a lake is likely sooner or later to find that the answer to some problem he is seeking to solve is to be sought in past history, and particularly in the way the lake was formed. Lakes and ponds have originated in many different ways and much ingenuity has been devoted to fitting them into schemes of classification. We shall not dwell on the groupings and subgroupings which have been suggested, but, in the first part of this chapter, prefer to take the lakes as they come, starting in the north of Scotland and travelling southwards.

The Great Glen is a tear in the earth's surface, and Loch Ness, which lies in it, provides an example of a lake associated with faulting. Loch Ness is 21¼ miles (35 km.) long and a little under one mile (1·6 km.) in average breadth, so it is a long and narrow lake; its greatest depth is 754 feet (230 m.) and its mean depth 433 feet (130 m.), so it is deep and steep-sided. Its mean depth is greater than that of any other British lake by quite a big margin, though there is one, Loch Morar, which is deeper at the deepest part (1017 feet = 310 metres). All these features are characteristic of tectonic lakes, that is lakes formed originally by movements of the earth's crust. Also in this class are some of the most striking lakes of the world, such as the Dead Sea and the lakes in the Great Rift Valley of Eastern Africa. These were caused partly by lateral tearing, as is the Great Glen, but there followed a lowering of a strip of the earth's crust so that what is now the floor of the rift valley was once level with the high land on either side.

Most of the other lochs in the Scottish Highlands owe their present form to the work of ice when the country was covered with it during the Ice Age and so were the llyns of the Welsh mountains and the lakes of the English Lake District. Indeed, nearly all the larger stretches of water in Britain were formed by glaciers, at least in part. In some cases their basins were gouged out by glaciers flowing down mountainsides, usually in valleys cut by a stream in an earlier, more clement period. When the mass of snow and ice and rubble reached the bottom of the slope it dug into the ground and excavated a great trench. This trench became the basin of the lake when the glacier retreated with the onset of warmer conditions. The mass of material which the glacier plucked from the land it passed over was deposited in mounds or moraines at the snout of the glacier, and some glacial lakes are dammed up by a moraine which makes them deeper than if they were contained in the actual excavation alone. Glacial lakes, like rift valley lakes, are usually long and narrow and relatively deep; Windermere, for example, is 10½ miles long but only half a mile wide on the average, and 219 feet deep at the deepest point. Their sides tend to be parallel and any major irregularity in the shore line is often of more recent age. Lakes of this type were formed only in hard rocks where the relief was rugged, and steep valleys concentrated the glacier and directed its excavating effort to a circumscribed area.

Also of glacial origin are the smaller lochans, tarns, and the small lakes in cirques, corries, or cwms which are often to be found near the tops of mountains. They are frequently circular in outline and they mark the place where the snow or ice piled up and a glacier took its origin.

An ice sheet covering a plain did not excavate because its effort was dispersed and not concentrated, but it did give rise to lakes none the less. As might be expected these are of a different type; Loch Leven is an example and no fisherman requires a biologist or geologist to tell him that there is something fundamentally different between Loch Leven and the Highland lochs. As the ice sheet which covered Scotland began to recede, a large lobe of the glacier flowing down the Forth Valley became isolated in the centre of the Kinross plain. It was surrounded by clay, stones, boulders, and suchlike products of ice erosion washed along in the water from the melting glaciers, much of it coming through a pass in the Ochils from the Tay Valley. A considerable depth of this material was deposited on the plain, but in the middle there was this big block of ice melting slowly because of its large size, like an iceberg in the North Atlantic. When it finally disappeared it left a hollow where it had been sitting and this filled up with water to become Loch Leven. The shape of Loch Leven is quite different from that of either glacier-cut or rift valley lakes: it is not much longer than it is broad, one axis being 3¾ miles (5·7 km.) and the other 2⅔ (4.1 km); its mean depth is only 15 feet (4·5 m.) and its greatest depth only 83 feet (25 m.).

The Cheshire Meres were formed in a similar way to Loch Leven, although subsidence of the land surface also played a part. Outside Britain there are many lakes of the same type: two groups, which are referred to in later chapters because they have been studied in much detail, are the numerous lakes in the Wisconsin area of North America and the Baltic lakes of Denmark, Germany, Poland and U.S.S.R.

Also characteristic of mountainous areas are the much smaller peat pools. These may occupy holes where stone for a wall or a house has been quarried or sometimes a rock basin of natural origin, but most of them are formed by the growth and then the erosion of peat. Some of the largest are to be seen on the Pennines, for the Pennines have flatter tops than the mountains in Scotland, Wales, or elsewhere in Britain, and it is on flat places that these pools develop. Vegetation, of which bog-moss (*Sphagnum*) is usually an important constituent, dies and accumulates over a long period of years, building up a bed of peat. At a certain stage, for reasons which are not at present understood, the peat becomes unstable, and hollows are eroded by the action of wind and rain. The surface becomes dotted with small pools and, as further erosion takes place, these coalesce. Finally a channel becomes eroded through the rim of the peat bed and all the water runs away. The building-up process then starts again.

Larger bodies of water on the Pennines are few, apart from man-made reservoirs which are now characteristic features of the landscape.

Lough Neagh in Northern Ireland is the largest sheet of fresh water in the British Isles, with a surface area of 153 square miles (393 km.²). It was formed in a way different from that of any of the other lakes so far encountered, and is volcanic in origin. There was no volcanic mountain like Etna or Fujiyama, but basaltic lava welled up from fissures in the ground. It flowed freely over the countryside and eventually solidified as a flat plate-like capping. Later it sagged in the middle and the depression so formed is Lough Neagh. The greatest depth is only 56 feet (17 m.), so it is even shallower than Loch Leven.

There remain to be explored the more recent geological formations of south-east England, and on them there are few large bodies of fresh water, though they are not on that account of any less interest to the freshwater naturalist. The best-known sheets of water are the Broads of East Anglia. The scientific mind, like Nature, abhors a vacuum and there was no dearth of armchair theories about how the Broads had been formed when Joyce Lambert, a botanist, and J. Jennings, a geographer, set out to collect some facts.

Dr Lambert pushed her way through the dense fens along straight lines from edge to edge and took borings at regular intervals. At many places the peat was of a different type at different levels, at others it was uniformly of a type that indicated accumulation at the bottom of standing water. In due course an elaborate and plausible explanation of the origin of the Broads was formulated and it might have remained the accepted one for a very long time, a great deal of hard work having gone into the collection of the evidence. However, Dr Lambert thought it prudent to continue her borings, and turned up evidence which demolished the theory. She found peat composed of different plant associations at different levels and peat that had accumulated uniformly in water so close together that the plane between them must be vertical; indeed there was evidence of columns of the former surrounded by the latter. There could be no explanation of this except excavation by human agency and the research became primarily historical.

No direct evidence has been found but the circumstantial evidence is convincing.

Documents of the thirteenth and earlier centuries refer to turbary rights in the region of the Broads, and there are records of much peat-cutting in parishes where there was no source of peat other than the fens where the Broads now are. There is no reference to water. After about 1350 there are few references in old documents to turbary, but frequent references to fisheries. There is, therefore, good reason to believe that the Broads are old peat-cuttings which became flooded between 1300 and 1350 probably as a result of some change in the relative levels of land and sea (Ellis, 1965).

A river tends to build up a deltaic plain at the end of its course and it inundates this plain every time it rises a little above its normal level. Parts of the plain will be under water only at the height of a flood, parts will be permanently marshy, and parts will be under water all the year round. This is the normal and accepted state of affairs in regions of the world where man has done little towards controlling and taming nature: the Rivers Tigris and Euphrates may be taken as an illustration (Fig. 6). In Britain, however, man has long since decreed that there is a place for everything and the place for water is within well defined banks; any breaking out and overflowing is an irregularity and often a catastrophe, and the victim of a flood is not consoled by the assurance that it is "natural".

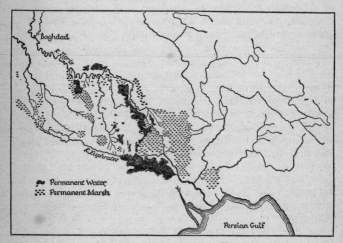

Fig. 6 Lower courses of the Rivers Tigris and Euphrates

The East Anglian fens originated when a flat clay-floored valley opening into the Wash was flooded by the sea after the Ice Age, owing to a slight lowering of the land level. Silt banks deposited by the sea gradually cut it off and it became a great inland lagoon. It was shallow, and rich in nutrient salts. Conditions were, therefore, good for plant growth and the resulting vegetation was luxurious. The dead remains accumulated and formed peat which filled up the lagoon rather rapidly, speaking in geological terms, till open water was left only in a few meres, which must have been very like the Broads today. Man coveted the rich soil and in the seventeenth century he successfully started drainage and reclamation. Now the meres have gone, the natural vegetation is to be found only in a few carefully tended preserves, and the fenland presents to the pond-hunter no more than an endless series of ditches, great and small.

Travelling a little farther south, we come to the chalk region; and a more waterless expanse than a chalk down cannot be found anywhere in the country. But even here there is something to interest the freshwater naturalist. Man has been wont to run stock over the downs for centuries and, in order to provide them with water, he has built ponds which have received the name of dewponds. There are few subjects about which more nonsense has been written. One explanation offered, even by people who should know better, is that dewponds are made by a secret process which insulates them from the surrounding land. When heat is lost at night by radiation from the surface, warmth from the lower layers is conducted upwards, and therefore the temperature at the surface does not reach a very low point; but this upward conduction cannot affect the dewpond because it is insulated. Accordingly, it is alleged, the dewpond area gets very cold, the atmosphere above it is chilled and moisture is deposited. The difficulty about this theory is that, if the dewpond were so effectively insulated from the land below, it would get very hot when the sun shone on it by day and much water would be lost by evaporation. Furthermore, considerations of the respective latent heats of water and chalk (that is the amount which a given volume of each would lose in a given time) have been ignored. Several people have examined the problem both experimentally and theoretically and the whole fallacy has been exposed more than once. Mr A. J. Pugsley (1939) has returned to the attack in a small book published by *Country Life*, but it would be optimistic to expect that the myth has been exploded. There is certainly a secret process in the making of dewponds and it has been handed down from father to son in certain families, but its aim is the construction of a waterproof bottom which will last for many years without cracking.

The dewpond, in effect, is a shallow pan of concrete or clay, and, though sometimes it is situated on top of a hill where it must rely entirely on rain, often it is located to take advantage of storm water, particularly where a road presents an impermeable surface. The belief that dewponds date back to the Neolithic Age is erroneous.

We have now worked our way down to the south of England and come to the coast to study a freshwater lake which owes its origin to sand-banks thrown up by the sea. To the west of Bournemouth lies Poole Harbour, a big enclosed area connected with the open sea by a small entrance, which cuts through a narrow strip of land and so makes two peninsulas. The one which lies to the west is known as South Haven Peninsula, and a conspicuous feature of it is the Little Sea, a shallow lake over 70 acres (28 ha.) in extent. Particular interest lies in the fact that the origin and development of Little Sea can be traced in detail from the information given on old charts. The first of these, dating from the reign of

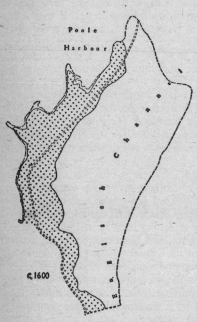

Fig. 7a Formation of the Little Sea, c. 1600

Henry VIII, is not very accurate, but from it and one or two later charts a fair deduction is that the peninsula then comprised only land of the Bagshot Sand formation, with a small more recent sandspit at the tip. This is shown in Figure 7a, but the sea and the area between tidemarks are omitted from this figure, as any attempt to include them would be based largely on conjecture. A chart of 1721 is remarkably accurate. It shows that a sandbank, thrown up parallel with the land existing in the previous century, has enclosed a lagoon which is apparently a sheet of water at high tide but at low tide an expanse of bare sand, except for water standing in drainage channels. There is a wide beach, shown stippled in Figure 7b, and a detached sandbank lying to the north of the channel draining the lagoon. Rather more than a century later, in 1849, a survey shows considerable development; a second ridge has

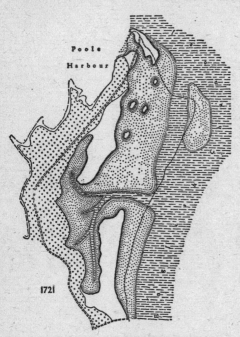

Poole Harbour

1721

Fig. 7b Formation of the Little Sea, *c.* 1721

been thrown up parallel to the first in the northern half and marshy area indicates the depression between the two; a third is foreshadowed by a long sandbank which now bounds the outflow on the seaward side; it is shown white in Figure 7c, the convention used to denote land above high water, though strictly it should be stippled as, according to the chart, it was covered by the highest tides. The sea runs in and out of the channel between this bank and the second ridge, and water apparently stands in the north and south portions of the lagoon at all stages of the tide. Today (Fig. 7d) there are three dunes, and the Little Sea is an inland lake with water which is actually soft and rather poor in dissolved salts. Other, smaller, bodies of water have come into being and the slacks between the dunes are extensively marshy; man-made cuts traversing them testify to an attempt at some earlier date to drain them, presumably to obtain pasturage.

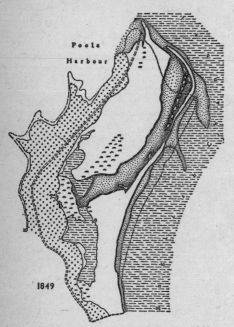

Fig. 7c Formation of the Little Sea, *c.* 1849

And so, thanks to the painstaking research of Captain C. Diver (1933), it is possible to reconstruct in detail the changes which brought Little Sea into existence. There are other sheets of fresh water of similar origin, but no one has pursued inquiries into their early history. Some have obviously been formed more simply, and Llyn Maelog and Llyn Coron in Anglesey, for example, lie in long transverse depressions which the sea has blocked at the ends with sand.

Wherever man has had available an impervious soil he has

tended to make ponds and lakes, to provide him or his animals with a water supply, for ornament or for sport. A favourite site is a narrow valley which can be flooded by the erection of a dam (Fig. 8), for building a dam is comparatively simple, while sufficient excavation to make a pond of reasonable size is a big and costly undertaking. Where there is hard impervious rock, fish-ponds are sometimes very numerous; for example, in the southern part of the Lake District the staff of the Freshwater

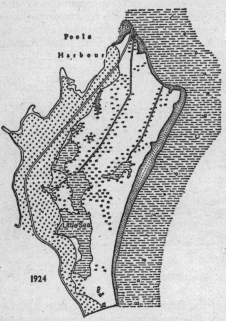

Fig. 7*d* Formation of the Little Sea, *c.* 1924

Biological Association have nearly fifty under observation within easy reach of their laboratory.

On heavy clay soil the farmer frequently digs a hole in every field in order to form a pond from which his animals may drink. Many other pieces of water are the by-products of man's activities. Quarrying for stone, or digging out clay for bricks, produces an impermeable basin which the rain will

ultimately fill. Excavating sand and gravel for railway ballast
and other purposes may extend down below the water-table so
that a pond results. Underground mines and tunnels sometimes cave in and cause at the surface a depression which fills
with water.

The prosperity of the fifties and sixties and the boom in
aquatic sports such as fishing and boating has meant that
many gravel pits that might otherwise have been used for the
disposal of rubbish have been saved as lakes. On the other
hand, many small ponds are disappearing, because, with state
aid to water supplies for farms, they are no longer necessary
for watering stock. Indeed their use for this purpose is actively
discouraged, since it has been shown that cattle contract
Johne's disease by drinking from fouled ponds.

Fig. 8 Hodson's Tarn, an artificial moorland fishpond

These are some of the main ways in which bodies of fresh
water have originated. There are others, less important in the
British Isles, but a catalogue of them would serve no useful
purpose here. Our main interest is with the plants and animals
of water, and the next stage is to notice how lakes may be
classified according to the biological processes going on within
them.

A lake receiving the drainage from rich cultivated land will
be 'productive', because of the nutrient salts it receives, that is,
a large quantity of plant and animal material will be produced
in the upper layers. Many of these organisms will decay in the

lower layer, which, if the lake is stratified, may become depleted of oxygen. A second condition is that the hypolimnion should be relatively small. A combination of good agricultural land and a shallow lake is typical of lowland country, and it is here that lakes with no oxygen in the hypolimnion generally occur. They are known as 'eutrophic'. An 'oligotrophic' lake, that is, one in which the hypolimnion contains oxygen, is typical of mountain conditions where the drainage area is unproductive and lakes often occupy deep basins. For many years the difference was thought to be fundamental, and an elaborate classification arose on a foundation which had been simple originally. As knowledge accumulated, it became evident that the distinction was not as basic as had once been thought, and it is no flight of fancy to say that the edifice of classification was brought crashing about the ears of the assembled company by Professor H.-J. Elster, in a masterly review at the International Congress of Limnology in 1956. Since then the tendency has been to study the primary productivity of a lake, that is the amount of algal material produced in the open water during a year, and to arrange the lakes in a series according to the figure obtained.

Shortly before the First World War, the late Professor W. H. Pearsall started a study of the Lake District lakes the basins of which were all formed in the Ice Age. Whether he was familiar with the continental ideas and ignored them, or whether he was not aware of them, we shall probably never know. Anyhow he arranged the lakes in a series with no attempt to delimit and define categories, although Esthwaite, at the productive end of the series is eutrophic, and Wastwater, Ennerdale, and Buttermere at the other are fine examples of an oligotrophic lake. This concept stimulated a great deal of work, and though Pearsall's original ideas have been modified, the basic soundness of the idea has been revealed by research in several fields. Pearsall noted that the unproductive lakes lie in the hard Borrowdale volcanic rocks right in the main mountain masses. Consequently the valley sides are steep, the area of flat valley bottom is small (Fig. 9) and rain falling on the drainage area will flow over much bare rock and scree. Consequently it bears little in solution when it enters the lake. The unproductive land supports no more than a farm or two, and few other than farmers have been tempted to settle in the restricted area available. This, however, has also been influenced by the re-

Fig. 9 Buttermere, an unproductive Lake District lake

Fig. 10 Esthwaite, a productive Lake District lake

moteness of the valleys which are distant from the main lines of communication.

Windermere and Esthwaite Water (Fig. 10) are the two most productive lakes. They lie in the south of the district in a zone of Bannisdale slates, which, though hard rocks, are softer than the Borrowdale Volcanics and have weathered more. Much of the drainage area is floored with the products of weathering and is relatively flat. Obviously rain-water seeping through soil will dissolve out more than water trickling over solid rock, and so the streams and rivers entering Esthwaite and Windermere bring with them a higher concentration of nutrient salts than those flowing into Ennerdale. But the flat land also attracts the farmer and the cultivator who seeks to improve the soil by adding manures to it. Some of these find their way into the lake, and so the difference between the two is enhanced. Within the last century Windermere, particularly, has become a residential resort. The result is that much human sewage enriches its waters and makes still greater its difference from Ennerdale.

Esthwaite Water is a relatively shallow lake and, as already stated, eutrophic.

Position in the series developed by Pearsall was based on three factors, first the percentage of the drainage area which is cultivated, second the percentage of the shallow water region which is rocky, and third the transparency of the water. The first two factors are fundamental; the third is partly fundamental and partly a result, because the transparency of the water depends both on the amount of mineral matter in suspension and on the quantity of life present, provided there are no extraneous factors like staining from peat or pollution by mine washings. In the Lake District none of the larger lakes except Bassenthwaite contain peat-stained water, and pollution from mine washings, though it does occur, is fortunately rare. Table 2 shows the Lake District lakes arranged according to these three factors. The figures in the last column show the depth at which a white disc, 7 cm. in diameter, could just be seen.

On the whole there is a serial increase or decrease in each of the three columns. The most notable anomaly is the low transparency of Bassenthwaite, occasioned by its being the only lake of which the water is stained with peat. The amount of light at different depths in Bassenthwaite, Windermere and Ennerdale is shown in Figure 11, expressed as a percentage of the intensity at the surface.

Table 2. The sequence of Lake District Lakes (Pearsall, 1921)

Lake	Percentage of catchment area cultivable	Percentage of lake shore to depth 30 feet rocky	Relative transparency of water (metres depth)
Wastwater	5·2	73	9·0
Ennerdale	5·4	66	8·3
Buttermere	6·0	50	8·0
Crummock	8·0	47	8·0
Hawes Water	7·7	25	5·8
Derwent Water	10·0	33	5·5
Ullswater	16·6	28	5·4
Bassenthwaite	29·4	29	2·2
Coniston	21·8	27	5·4
Windermere	29·4	28	5·5
Esthwaite	45·4	12	3·1

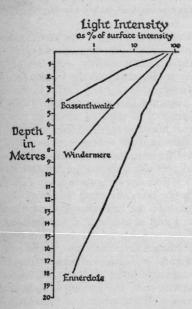

Fig. 11 Penetration of light into three Lake District lakes

Work on cores, started just before the war, had as its original aim the elucidation of the history of the lakes, but, like many another new line in research, an essentially opportunist activity, it proved most fruitful in a line other than the one aimed at and it revealed more about the land than about the water. However, as a lake is strongly influenced by events in the drainage area, the findings are relevant. Most animals disappear completely, but the shells of some waterfleas (Cladocera) and the heads of some chironomids do not decompose and persist in the cores. Similarly many algae leave no trace, but the siliceous skeletons of diatoms (e.g. *Asterionella*) that have lain in the mud for thousands of years are still identifiable. In contrast the pollen of nearly every plant that produces any does not decompose and, since that of almost all species is distinct, an examination of cores gives a picture of what the land flora was like when the particular layer of mud under examination was deposited. Research on pollen in cores from bogs and other places where soil has been accumulating since the Ice Age was in vogue all over Europe at the time, which was fortunate, because events in the lake cores could be related to events elsewhere, and some of these had been dated by one means or another. Chemical analysis of cores also yielded a large amount of information about the past.

The lower part of a core from Windermere consists of clay which, on examination, proves to be made up of alternating layers of very fine and coarser particles; it is accordingly referred to as laminated clay by Dr Winifred Pennington, who has described the cores. Above the laminated clay, which is pink, lies a grey layer and above that more pink clay, which may or may not be laminated. On top of this is a thick column of brown mud which extends nearly to the surface; it is capped by a fourth kind of soil, a black deposit which Pennington refers to as ooze.

The laminated clay contains very little organic matter and few remains of plants, and was almost certainly formed towards the end of a glacial period, for similar deposits are being laid down today in certain glacier-fed lakes. During the summer both coarse and very fine particles are washed into the lake. The coarse particles sink almost immediately but the fine particles remain in suspension for a long time. When winter comes the inflowing streams freeze, and so no particulate matter is brought into the lake, but fine particles left over from the summer are still settling. The result is a summer layer of coarse and fine particles and a winter layer of fine particles only.

The low organic content of this deposit and the scarcity of plant remains indicate that there was little life in the lake when the laminated clay was being laid down. In contrast the grey mud contains the remains of animals and plants, and the lake was evidently more densely populated during the period when it was being laid down. These organisms were associated with an improvement in the climate, which is known as the Allerød period, because it was first discovered near the place of that name. It was followed by a return of glacial conditions when the pink clay with few remains of organisms was laid down again. Professor H. Godwin has had dated by means of the C^{14} method, a technique which will no doubt be more widely used when facilities for it are more widely and more easily available, a sample from the Allerød layer and found it to be some 12,000 years old. The ooze at the top in which *Asterionella* suddenly becomes common obviously represents enrichment of some kind. Today the effluents from the sewage works are probably the main sources of enrichment, but this is recent. The population, particularly the holiday-making one, has been increasing since the railway came to Windermere in 1847, but the transition from earth closets to running water sanitation has been slower, and it is doubtful if the lake was being seriously enriched from this source a century ago. If mud is being deposited on the bottom of Windermere at a rate that has been constant over the last few hundred years, whatever caused *Asterionella* to appear happened about two centuries ago. It therefore probably antedates the tourist completely, and is to be sought in improved agricultural practice of which there is some evidence early in the eighteenth century.

Mackereth's conclusions from chemical analyses upset certain ideas of long standing. The lakes were richest chemically in their earliest days, when the land was covered with rock fragmented by the ice and exposing a great area of surface from which the rain could leach nutrient salts. Esthwaite was eutrophic at an early stage and presumably, therefore, the lakes were richest biologically when they were richest chemically. Some of the algal species identified in the cores support this view. The rocks are hard and when a fresh surface has been leached for a time water dissolves little from it. The lakes slowly became less productive. A climatic change and the arrival of man, who started to fell trees, resulted in more erosion and enriched the lakes, but as more stable conditions were established production fell again. A thousand years ago the Norsemen arrived, and since then man has been the most important agent affecting the lakes. The increased production in some lakes on account of their situation has already been described.

5. Rivers

It is possible to entitle a chapter 'a typical lake' and to fill it with an account of physical and chemical changes which follow an annual cycle in a great many lakes. Another chapter is entitled 'different kinds of lakes', though here a certain degree of accuracy has been sacrificed to obtain a short title, and the account includes pieces of water that are too small to be regarded as lakes in the strict sense. In neither chapter is there much about fauna and flora, the plan being to describe freshwater animals and plants in chapter 6 and then to pass to an account of the various communities found in different biotopes. A biotope is a region in which the conditions are of such uniformity that the plant and animal communities do not vary much; the stony substratum of a lake, weed-beds and the open water are examples of biotopes. The purist would prefer the term biocoenosis for the assemblage of animals and plants that inhabit a biotope, but here the more general term 'community' is used.

It is impossible to treat rivers in the same way. Lakes are all recent in geological terms, and many of what were called lakes in the preceding chapter are recent in historical terms, having been made by man. Each lake was formed by one event taking place in a limited area. Water courses are much older and have continued their existence in spite of the events which formed lakes. For example the rivers of the English Lake District tend to rise near the middle of the area and radiate from it. This pattern was presumably established at a time when the mountains exposed today were covered by a dome of younger rocks. The watercourses cut down through this and later eroded the underlying rocks in the same direction, though these originally had the form of a ridge not a dome. The disappearance of the covering layers left the old rocks scarred by valleys whose direction bears no relation to the way in which they were laid down. Each watercourse is, therefore, modified today by many geological strata.

The plan of this chapter is to describe the few rivers about which something is known; how far they are 'typical' only further work will show. It has also been found necessary to include some information about the plants, and about human activities. Whereas these have brought whole new bodies of standing water into existence, they have modified rivers. These modifications, however, have been extensive and have affected every British river of large size. Waste disposal, water supply, water power, drainage and navigation have been the main activities through which man has altered water-courses.

The history of investigation of rivers fits more easily into the account of their misuse in Chapter 14. It suffices to record here that important surveys were carried out between the wars by the late Mr F. T. K. Pentelow, Dr R. W. Butcher and colleagues in the Ministry of Agriculture team. Since the war knowledge about small stony streams has advanced greatly, investigations having been made in most of the upland areas of England, Wales and Scotland. One Lake District river has been investigated by Drs R. Kuehne and W. Minshall working in England during the tenure of a year's fellowship from America. Of recent years the number of biologists on the staffs of the River Authorities has increased considerably. It seems to be envisaged, however, that their task is to analyse as many samples as possible from as many stations as possible in order to keep a check on the condition of the river and its tributaries. For this purpose it is often reckoned that identification to species is not necessary. For basic information about the lower courses of rivers and their inhabitants, we still have to turn to the old surveys.

The property of water important to the study of lakes is its density at different temperatures; for the study of rivers, its flow downhill. This has a direct influence on organisms that live in it and a possibly more important indirect one through its effect on the substratum. Table 3 is taken from Tansley (1939) but its original source is a text-book on river and canal engineering. An engineer contrives even gradients and neatly regulated bends; Nature does not. The irregularity of a natural water-course produces a mozaic that confounds the systematic mind at the start and ensures that any scheme of classification is no more than a rough general guide. What does happen in nature? In attempting to answer that question, we shall take an imaginary river, but admit at once that our imaginations owe much to familiarity with the Lake District. The rocks there are hard, impermeable and often steep. In places water flows a long distance over a flat sheet of rock, but generally it has eroded a gulley of some kind. In this rapids generally alternate with pools in parts of which conditions are surprisingly quiet, especially if the stones and boulders are large. In the rapids large stones and boulders tend to jam in the gulley and hold up a bottom that is far more stable than might be expected on so steep a gradient. Conditions in this zone must obviously depend on the relation between the dip of the strata and the angle at which they are exposed, and on the size and form of the fragments which break off the rock. Where the gradient becomes less steep, moderate-sized stones plucked

Table 3. Relation of current speed and nature of river bed

Velocity of current per second	Nature of bed	Habitat
More than 4 ft. (1·21 m.)	rock	torrential
„ „ 3 ft. (0·91 m.)	heavy shingle	torrential
„ „ 2 ft. (0·60 m.)	light shingle	non-silted
„ „ 1 ft. (0·30 m.)	gravel	partly silted
„ „ 8 in. (0·20 m.)	sand	partly silted
„ „ 5 in. (0·12 m.)	silt	silted
Less than 5 in. (12 cm.)	mud	pond-like

from the rocks above and washed down begin to come to rest. In the Lake District they tend to be flat and accordingly they have an inherent stability. Further breaking-up is taking place all the time and as the smaller pieces are rolled downstream their edges are rounded. This produces the unstable bottom of round stones that is often found some distance down the valley, if the gradient falls evenly. An outcrop of rock is a fresh source of flat stones, and it, or any other obstruction, produces a striking alteration of the flow pattern, confining swift current to the surface layers. Settling of gravel, sand and finer particles becomes possible, and, as these fill up the interstices between the larger stones, they produce a remarkably stable bottom. This happens also during a spell of low water. Percival and Whitehead refer to it as a 'cemented' bottom. It is one which occurs almost everywhere, but generally it is covered by loose stones. Occasionally some new obstruction halts the downward flow of loose stones and then the stream is floored by substratum of this type.

Flowing onward, the river generally comes to a plain, often one of its own creating, the gradient approaches nearer and nearer to the horizontal and flow decreases. First gravel, then sand and finally silt settle to the bottom and provide a substratum in which plants can take root.

Probably few except those charged with the task of dredging it realize how much material is being deposited on river beds. It is not a continuous process and varies greatly with intensity of rainfall. At intervals exceptional downpours, often restricted to a comparatively small area of the mountains, make considerable alterations to a river bed, which may remain comparatively unchanged until the next downpour. But change never wholly stops; a boulder may stabilize a stretch for many years but all the time it is being chipped away by the smaller stones washed past it until the day must come when it is no longer large enough to withstand a flood. Away it goes and a considerable section of the adjacent bottom with it until a new pattern is established.

For the biologist the important distinction is between the upland reaches, where erosion is taking place, and only plants, such as mosses, that can attach themselves to flat hard surfaces provide cover for animals, and the lowland reaches where a plain is being built (or would be if the drainage engineers permitted) and rooted vegetation grows. Dudley Stamp (1946) recognizes three zones; mountain, foothill, and plain. Butcher has proposed a classification of rivers according to which of these zones they rise in. In mountain areas with hard rocks the rivers will traverse all three zones, but where there is chalk or other pervious rock the river may spring from the foothill or the plain region. His scheme, however, has not caught on, and most workers agree that an entire river may be so diverse that it will not fit with other rivers into a category. Schemes for recognizing zones within a river have, in contrast, been popular. The best known goes back a long way and has been elaborated in recent years especially by fishery workers. It is based on the species of fish found, which appears to have a fair correlation with the slope. One drawback is that some of the fish do not have a wide geographical distribution. Dr Kathleen Carpenter (1928) has adapted it for British waters:

1. The Headstreams and Highland Brooks are small, often torrential, and without fish. Temperature conditions vary greatly. Low temperature is common, but a stream that arises from shallow soil may be warm, and a slow-flowing stretch may soon reach a high temperature on a sunny day.

2. The Troutbeck is larger and more constant than the headstream. Torrential conditions are typical and the bottom is

composed of solid rock, stones, and boulders, with perhaps some gravel. The trout is the only permanent fish of the open water though the miller's thumb (*Cottus gobio*) is found sheltering among stones. These first two zones together correspond roughly with Stamp's upper or mountain course.

3. The Minnow Reach is still fairly swift and patches of silt and mud are only to be found in a few places protected from the current, but higher plants, notably the water crowfoot, *Ranunculus fluitans*, are able to gain a foothold. This is roughly the middle course of Stamp. It is the *Thymallus* (grayling) zone of the continental workers, but Carpenter rejects this name because the grayling is not a widespread species in Britain.

4. The Lowland Reach is slow and meandering, with a muddy bottom and plenty of vegetation. Coarse fish are characteristic, and on the continent of Europe it is known as the bream zone.

Tansley classifies rivers into five zones, basing his system very largely on the work of Butcher (1933).

Zone 1 is described as very rapid. Where vegetation is present at all, the important plants are mosses and liverworts; higher plants are often absent altogether and never dominant. This class includes all Carpenter's headstreams and highland brooks and part, at least, of the troutbeck.

Zone 2 is moderately swift with a bottom of stones and boulders, but with occasional patches of finer material in which a small number of higher plants can gain a foothold. *Ranunculus fluitans* (or sometimes *R. pseudofluitans*), the water crowfoot, is the most important.

Zone 3 has a moderate current with a gravelly bed. The list of higher plants is much longer. The water crowfoot is still the most important, others are the simple bur-reed, *Sparganium simplex*, several species of *Potamogeton*, and the Canadian pond-weed, *Elodea canadensis*.

Zones 4 and 5 are medium to slow, and very slow or negligible respectively. The list of higher plants is long and, as it varies a good deal from river to river, confusion rather than clarification would be the result of reproducing it here; but it may be noted the water crowfoot is usually *not* an important constituent. It is impossible to equate this classification exactly with that of Carpenter, but Zone 2 and part, at least, of Zone 3, correspond with her minnow reach, and Zones 4 and 5 and perhaps part of Zone 3 correspond with her lowland coarse fish reach.

Against this background a few British rivers which have been studied in detail may now be examined. The Lake District, as was described earlier, is drained by rivers which radiate from the centre. Their valleys were enlarged by glaciers during the Ice Age and generally deepened in such a way that a lake was left when the ice retreated. The Duddon is one of the few valleys in which there is no lake. Its highest tributary, Gaitscale Gill, rises at an altitude of 735 m. (2400 ft.) in a flat area covered with bog and small pools of open water. Beyond this it tumbles steeply down the fellside, dropping 300 m. in 900 m. (33%). At the foot of this slope there is a delta of large stones under which the water disappears in dry periods. Exceptional rainfall towards the end of the year during which Kuehne and Minshall were at work enlarged this delta, and incidentally carried away every one of about twelve maximum and minimum thermometers which they had buried in various parts of the system. The analyses made by these two workers showed that the calcium ranged from 0·57 to 1·10 parts per million in Gaitscale Gill, a low value, even for the Lake District, but after the flood the concentration below the delta rose to 3·0 p.p.m. This illustrates the point, stressed earlier, that freshly exposed faces yield more nutrients than those which have been leached for some time. The highest temperature recorded in the gill was 17·2° C., 5·6° C. lower than the highest temperature recorded elsewhere in the system.

Several streams run down the fellside parallel with Gaitscale Gill and feed the main river which here runs roughly westwards. Beside it runs a road, originally made by the Romans, probably as a line of communication through the area to facilitate the subjection of the natives (Rollinson, 1967). It is now used mainly by tourists, for there are no dwellings beside it between Langdale at one end and Eskdale at the other. The river swings round to take a southerly direction in an upper valley with a comparatively slight incline. There were four farms in this valley, but only two are used for farming today. In autumn a few green fields around them stand out among the predominant greyish-yellow of the poorer pastures, probably as a result of liming. Sheep, which range far and wide over the surrounding fells, particularly in summer, are the chief product. A few conifers, planted recently, are the only trees.

The slope steepens to separate the lower from the upper

valley. The slope in the lower valley is 1%, the river is floored
with round stones of all sizes, and it flows torrentially down to
the estuary. There is no plain region, and only the first two of
Carpenter's four zones and the first of Butcher's five can be
recognized. There are two small villages in the lower valley,
residences scattered outside them, and some twelve farms on
which cattle as well as sheep are reared. Deciduous woods
cover extensive areas of the valley sides. The upper valley
comes to an end about 175 m. above sea level, the lower at sea
level. The difference in climate between the two is obviously
great but no figures are available. The temperature of the swift
river gives no indication of it. The valley walls rise steeply on
both sides but to the west there is a plateau over which flows
the longest tributary.

The River Duddon is about 11 miles (18 km.) long. We pass
from it to the River Tees (Butcher, Longwell, and Pentelow,
1937) which is about 100 miles (160 km.) long. It rises in the
Pennines and flows to the North Sea. The gathering ground
drained by the headstreams is fairly large. It is high above sea
level, it receives a relatively heavy rainfall of about 60 inches
(1520 mm.) a year, and the rock is impermeable. The result of
these four factors is a severe scour in time of flood, and this
has carved out a deep bed. Consequently the energy of a flood
is not dissipated in inundating the surrounding country and
the effect is concentrated on the river-bed. On one occasion
some carts were being filled with gravel at the water's edge
when the river rose so suddenly that the carts had to be
abandoned and were swept away. This illustrates a most important condition affecting the plants and animals of the river.

The small town of Croft is about 45 miles (70 km.) from the
mouth and about 65 miles (100 km.) from the source, travelling
by river, and is a convenient dividing point. Above Croft all the
river is rapid with little rooted vegetation and few fish except
trout, grayling, and minnows. The river downstream is still
moderately swift but there is much more rooted vegetation and
various coarse fish are plentiful. After flowing for 20 miles (32
km.) from Croft the river reaches the head of the estuary,
which is some 25 miles (40 km.) long.

The Tees rises on Cross Fell at a point about 2,500 feet (760
m.) above sea level. Many small tributaries also rise just on the
eastern side of the Pennine watershed, and some of them
originate in thick peat beds. One, for example, drains the peat
pools which were mentioned in Chapter 4. These little streams

run down the hillside with a fairly though not extremely rapid flow, because the eastern slopes of the Pennines are not steep. About six miles from the source three main tributaries have coalesced and the river is some 12 metres wide with a fair flow over a bottom of stones and boulders. Then it enters a quiet stretch and for three miles the current is sufficiently slow to allow the deposition of some fine sediment, which provides a foothold for a few higher plants, *Potamogeton alpinus*, the alpine pond-weed, *Callitriche intermedia*, the water starwort, and *Sparganium simplex*, the simple bur-reed. The only other attached plants are the mosses, *Fontinalis antipyretica*, and *Eurhynchium rusciforme*, and, at certain seasons, the algae *Lemanea fluviatilis* and *Cladophora glomerata*.

This slow stretch provides a pretty example of the sort of exception to the general plan which is to be found in almost any river. It is caused by a stratum of hard rock, and at the end of it there is a fine waterfall. A little farther on, some 16 miles (20 km.) from the source and 1,000 feet (305 m.) above sea level, the river strikes a road and some human habitations, and comes to the end of what may conveniently be taken as the first part of its course. A chemical station was set up just here and the results obtained are shown in Table 4.

Table 4. Some dissolved substances in the Tees near the end of the first part, that is above any pollution

	Maximum value	Minimum value
Calcium carbonate	57	12 parts per million
Calcium oxide	33	5 ,, ,, ,,
Magnesium oxide	8	0 ,, ,, ,,
Sulphate	12	2 ,, ,, ,,
Chloride	10	5 ,, ,, ,,
Dissolved oxygen	95	86% of saturation value

The average amount of calcium is about 12 parts per million, and so the water is soft although the river has been flowing over limestone. But there are big fluctuations in the concentration of all the substances except oxygen, which is plentiful at all seasons and at all times of day and night.

Trout occur in this part of the river and may be taken almost up to the source. They are plentiful but of small size, the average length being but six inches (15 cm.).

The next part of the course extends all the way down to Croft, which is at the point where the river changes in character. The river crosses several geological formations, which affect its nature, but it remains rapid throughout with a bottom of bare rock or stones. There is hardly any rooted vegetation.

The main difference between this and the preceding part of the river is that it receives sewage effluents from towns along its route. The first place of any size is Middleton-in-Teesdale, some twenty-two miles from the source, and the biggest is Barnard Castle, about eight miles farther on. The population connected with the sewage systems of the two places was 1,700 and 5,000 respectively when the survey was made. Below the outfalls there were changes in the flora, and there can be no doubt that these changes were directly attributable to sewage and the products of sewage decomposition. Below some of the larger works the dominant organism on the river bed was the sewage fungus. Father downstream there was a characteristic association of algae, but this finally gave place to the same association as was found in the upper waters, where there was no pollution. Below the smaller sewage outfalls there was no sewage fungus, but there was the characteristic change in the algal community encrusting the stones and boulders. In all this part of the river the amount of sewage was small compared with the volume of water into which it was discharged and pollution was not great. Oxygen concentration in the summer was lower than in the first part of the river (Table 4) but it never reached a seriously low level. Chloride rose, probably as a result of the sewage. The amount of calcium also increased and an average figure just above Croft was 24 parts per million; this rise was probably mainly due to the limestone over which the water had flowed.

Trout occurred throughout this part of the river and reached a greater size than in the first part, $\frac{1}{4}$ to $\frac{3}{4}$ lb. (120–360 g.) in weight with a few specimens of 3–4 lb. (1·4–1·8 kg.). It contained minnows almost throughout, and grayling in the lower half; thus, although there are no marked physical changes, the river enters the third of Carpenter's zones in this part of its course.

At Croft, 100 feet (30 m.) above sea level, there are several important changes in the river. It enters a great clay plain, laid down during the Ice Age, and flows across this in a meandering course, though with a fair flow. The stretch is too fast for a typical lowland course judged on purely physical grounds, and it is probably in the third of Butcher's five zones, for *Ranunculus fluitans* is one of the commonest plants; but it is in the last of Carpenter's four classes since coarse fish abound. Other changes are due to the confluence of the River Skerne, a large tributary, which is more calcareous and much more heavily polluted than the Tees. These three factors, different kind of bed, more calcium, and more sewage products, all influence the biology of the river below Croft, but it is not possible to measure exactly how great a part each one plays.

Most of the bed of the river is of medium-sized stones and gravel but there are occasional patches of sand. In water shallower than five feet typical plants are the water crowfoot, *Ranunculus fluitans*, and various species of pondweed, *Potamogeton*. These plants can colonize the gravel and sand, and when they have formed a large patch they cause a stagnant area on the downstream side. Silt settles here and accumulates rapidly if there is heavy pollution upstream. It is colonized by such plants as *Nitella*, the stonewort, *Elodea*, the Canadian pondweed, and *Potamogeton crispus*, the curly pondweed.

There was usually sufficient water in the Skerne to dilute the sewage it received to below the danger point, but in one summer there was a long hot dry spell as a result of which all the oxygen was used up, and the toxic products of decomposition without oxygen were liberated into the water. The Skerne itself had a thick coat of sewage fungus on its bed, and this organism extended for some distance down the Tees below the confluence of the two rivers. Its range varied widely according to the season of the year. In winter when, owing to the low temperature, the rate of decomposition of sewage is slow, it extended a long way downstream from the mouth of the Skerne, but in summer, when decomposition is more rapid, its range was less.

A green filamentous alga, *Cladophora glomerata*, the Blanket Weed, abounds where nutrients are plentiful, as they are below a sewage outfall where the organic matter has undergone the initial stages of decomposition. It appeared in the Tees towards the end of May and grew rapidly to form a thick carpet in the shallow water a long way down the river from Croft. Then the first flood in July would usually sweep it all away, and it would be seen no more until the following year. If it lasted long enough, it trapped a deposit of silt and enabled rooted plants to grow in places where otherwise the flow was too fast. Before the estuary was reached the algal community

typical of the upper, unpolluted reaches had become re-established on the stones.

Nitrogenous compounds and other products of decomposition were brought into the Tees by the Skerne, and the calcium
concentration was increased to some 30 parts per million.
There was less oxygen in this stretch during the summer than
there was farther upstream, and the lowest value was reached
during the time when the development of *Cladophora* was at
its height. The dense growth of this plant, respiring in the
hours of darkness, used up much of the oxygen and reduced
the concentration to between 50 and 60% of the saturation
value. This is well above the point at which deleterious effects
on fish are likely, and trout flourished in this, the last freshwater reach of the Tees, not uncommonly attaining a size of
1–1½ lb. (·45–·75 kg.). Coarse fish, chiefly dace and chub, were
abundant, and fishing was a popular pastime.

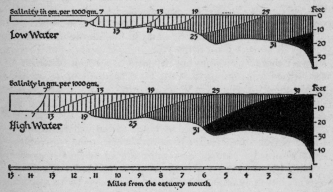

Fig. 12 Longitudinal section of the Tees estuary showing the salinity
at high and low tide

In the estuary, surveyed by Alexander, Southgate and Bassindale (1935), the most important natural phenomenon is the
salinity. The fresh water tends to float on the sea-water and the
result is a marked stratification. Figure 12 shows the average
conditions at high tide and at low tide, but it gives rather a
distorted picture because it is necessary to use such different
scales. Horizontally an inch represents about three miles, but
vertically it represents only about fifty feet. The surface current of fresh water draws up some water of higher salinity

from below it, and to replace this there is an upstream creep of water of high salinity along the bottom. The whole mass moves up and down with the tide as shown in the figure. It is estimated that the mean time for all layers of a body of water to pass through the estuary is about six days in dry weather, decreasing to about two and a half under average winter conditions.

The estuary has been much changed by the hand of man, and it must be admitted with regret that the Tees is typical rather than otherwise of larger British estuaries. From about midway nearly to the sea there is an extensive industrial conurbation. This requires a navigable channel so that its products may be removed by sea, and accordingly the natural tendency of the river to drop silt where it is checked by its meeting with the sea is counteracted by the continual activities of dredgers. The river is a convenient main drain and, at the time of the survey, the sewage from rather more than a quarter of a million people was discharged into it untreated. So were a variety of industrial waste products, of which the most important were tar acids and cyanides. Both these decompose gradually in the water.

Much water is taken in to cool condensers and machinery, and this results in a slight rise in the temperature of the estuary. Oxygen, it need hardly be said, is not plentiful in solution in the water. The amount used up depends on the temperature and also on the salinity, being greatest at salinities of between 15 and 25 parts per thousand. The lowest concentration of dissolved oxygen recorded during the survey was 9% of saturation.

The curly pondweed, *Potamogeton crispus*, the starwort, *Callitriche stagnalis*, and the two mosses, *Fontinalis antipyretica* and *Eurhynchium rusciforme*, which are abundant throughout almost the whole length of the freshwater part of the river, penetrate a little way into the brackish water. A few seaweeds penetrate a short distance from the sea but only four extend beyond the fringe of the brackish water region. *Fucus vesiculosus*, one of the brown bladder wracks, extends to beyond the middle point of the estuary, growing on wharves and piles between tidemarks; and three species of filamentous green algae occur throughout the brackish region.

It is difficult to determine exactly which fish dwell permanently in the estuary, as so many of the species recorded are migrants passing through, or casual invaders, but the three-spined stickleback appears to be a regular inhabitant, extending down to at least the upper reaches of the polluted part. The effect of the pollution on the fish, particularly the regular migrants, and on the lower animals is described in Chapter 14.

Reviewing the River Tees in the light of the classifications put forward at the beginning of the chapter, we find that it includes all of Carpenter's classes, for the lowest reach, immediately above the estuary, is dominated by coarse fish. On the other hand the last two classes, numbers 4 and 5, of Butcher's botanical classification are not represented, for the current is nowhere so sluggish that the water crowfoot ceases to be the dominant plant.

A contrast to the Tees is provided by the south country rivers rising in the chalk downs. Butcher has surveyed the plants of the Itchen, and there was a fisheries research station on the nearby Avon for several years before the war. Much of the gathering ground is chalk down. Rain falling on this sinks in and percolates relatively slowly so that it may not reach a hill-foot spring for months. The effect of heavy rain is, therefore, dissipated and it will not produce a marked flood wave as in the Tees. The other effect of the chalk is, of course, to render the water highly calcareous, and Butcher quotes a figure of 92 parts per million of calcium in the River Itchen.

Then the slope is not so steep. Moon and Green (1940) give a profile of the Avon and show that between Christchurch, which is at the mouth, and Salisbury the fall is about 150 feet in 39 miles, which is a little less than 4 feet per mile (0·075%). The river rises some 20 miles from Salisbury at an altitude of about 350 feet, so this upper reach, for which we have not been able to find accurate data, is somewhat steeper, and the figure for the whole river will be greater, but still far below that of 30 feet per mile (0.57%) for the Tees.

The springs giving rise to the Avon headwaters are usually at the foot of the chalk and often flow in wide valleys floored with gravel. Sometimes the streams have been broadened so that they flow over wide areas in which water-cress is cultivated. In dry weather the water-table often sinks below the surface of the gravel covering the impermeable stratum which is the true valley floor and the stream disappears. Sometimes, owing to the time which rain takes to seep through the chalk, there may be a long interval before the effect of a dry spell or a wet spell is manifest in the river.

Below Salisbury the Avon has been put to a variety of uses by man. One of the characteristic features is water meadows, although the method of farming under which they were engineered is now obsolete. The principle is to take water from the river in a main canal, which can be filled by the manipulation of sluices across its mouth and a barrage across the river. From this main canal the water is led into many subsidiary channels, from which it eventually runs over the land. It is gathered up in a complementary series of collecting channels and led back to the river at a lower level. The advantage of this system was that the grass could be watered at certain critical times of the year, and the farmer was independent of the capricious rainfall of this country. The significance of water meadows in the economy of the river today is that a great deal of flood-water finds its way on to them and runs back to the river slowly. This is a second reason why the effect of flooding is much less fierce in the Avon than in the Tees.

Dams and weirs are thrown across the Avon not only to deflect water for irrigation purposes but also to pen up a head of water to provide power for mills. Weirs and side channels to take excess water when the level of the river is high are usually to be found in connection with mills, and the result is that the river does not flow in a simple single channel but in a maze of anastomosing channels.

The water is rich in nutrient salts and, since there is no great scouring by floods, the rivers flowing from the chalk are heavily overgrown with a variety of aquatic plants. Butcher records that the commonest plants of the River Itchen are: *Ranunculus pseudofluitans*, water crowfoot, *Sium erectum*, the lesser water parsnip, and *Apium nodiflorum*, where the current is fastest; *Hippuris vulgaris*, the marestail and *Sparganium simplex*, the simple bur-reed, where it is somewhat less rapid; and *Elodea canadensis*, the Canadian pondweed, and *Callitriche stagnalis*, starwort, in the slowest reaches. The vegetation forms such thick beds that it has to be cut and removed to let the water pass, and also to make fishing possible.

Besides the game fish, for which these rivers are famous, there is a plentiful and varied population of coarse fish.

The Avon has no torrential head-stream region nor a typical meandering lowland reach. The whole river occupies a place somewhere in between these two, but it cannot be made to fit exactly into any of the various schemes of classification. There is no steady loss of gradient from source to mouth, as there is in the theoretical river, but a mosaic of faster and slower reaches due to the various artificial obstructions which man has thrown across the river.

A third river worthy of notice is the Lark, another of those surveyed by the Ministry of Agriculture and Fisheries team in connection with pollution (Butcher, Pentelow and Woodley, 1931). It is a small river rising in the East Anglian heights and flowing in a west by north direction to join the Ouse. The water is highly calcareous like that of the River Avon.

Only a comparatively small portion was surveyed, but this stretch is fraught with interest because it illustrates yet another effect of human interference. It may be remarked here that no south country river of any size is in a 'natural' state, and any account of it must dwell at some length on the modifications imposed by man.

The River Lark was once navigable as far up as Bury St Edmunds, though the last few miles were kept open with difficulty because the gradient was rather steep and the amount of water available was small. Eventually river traffic ceased to pay, and the locks fell into disuse. They are now derelict and the river flows in a bed which, having been widened to take barges, is too large for the volume of water which flows down it. This disproportion is particularly marked in one stretch which is now heavily overgrown with two emergent reeds, *Glyceria aquatica*, reed poa, and *Sparganium erectum*, the branched bur-reed. These plants probably established themselves first on beds of silt in shallow water. Their gradual spread would impede the current still more and result in further deposition of silt, and the process has continued and was still active at the time when the survey was made. The dense growth of reeds tended to deflect the current to the side, where it encountered and eroded a soft sandy bank, and so made yet bigger the area in which conditions were suitable for reeds. A stage had been reached where, when the reeds began to grow up early in the summer, above them the river flooded even though there had been no unusual rainfall, and below them a miller was hard put to it to obtain sufficient head of water to drive his mill.

Beyond this stretch overgrown with reeds there is a stretch overgrown with submerged pondweeds. In parts of it the current is sufficiently strong to keep a gravel bottom clear of silt and the water crowfoot is the dominant plant. Elsewhere the current is sluggish, the bottom is muddy, and the chief plant is usually *Potamogeton lucens*, the shining pondweed. In some places it is replaced by a community in which *Sparganium simplex* and *Sagittaria sagittifolia*, the arrowhead, are the dominant species. There was no evident difference in the river to account for these two distinct communities and at first they provided something of a puzzle. But a study of the activities of the human beings interested in the river at length provided the clue, and it was noticed that the bur-reed-arrowhead community was found in those parts where weed-cutting was most frequent.

Finally the river runs through a stretch of fenland before joining the Ouse, but unfortunately the survey stopped at the head of this stretch. The fenland river offers the extreme example of the lowland course. Left to itself it would follow a tortuous channel beset with marshes and stretches of open water. Changes of course might occur and the stream might split up and lose its identity in a number of small channels as does the Euphrates today. Figure 6 (p. 62) shows the lower part of the Euphrates and the Tigris, and gives a good picture of a lowland course which has hardly been interfered with. In Britain no fenland river is left to itself. The fen soil is valuable for cultivating and the rivers are· important as the means whereby the water pumped up out of the fens is got rid of. Vegetation, which would impede the flow of water, is removed and the channels are constantly dredged. Flood-banks are raised on either side, often at some distance from the river's brim, so that an expansion in width is possible when the river rises above its natural banks. Water left behind by a flood stands for a long period in this land between the river and the flood-bank, and the resulting 'washes' are characteristic features of the fenland landscape.

Most waterways were not created by man, though he has modified some of them considerably, but there is one group that owes its existence to human effort – the canals. The Exeter ship canal was built in the sixteenth century and a few artificial waterways persisting to this day date from even earlier times. But the title of 'father of inland navigation' is usually bestowed upon the third Duke of Bridgwater, at whose instigation a canal from Worsley to Manchester was built and opened in 1761. The commercial possibilities of this new means of transport were quickly exploited, and in the next forty years nearly 4,000 miles of canal were put into operation. After about 1800 the activity began to wane as the challenge from rail and road became ever greater. Today some of the canals have disappeared and others, though still containing water, are no longer used.

Even a used canal is surprisingly rich in animal life and an unused one is highly productive. Canals are almost confined to the lowlands and so their water is usually hard and rich in nutrient salts. There is sufficient flow to keep these replenished; but there is no danger of excessive flow after heavy rainfall of the kind which may wash away so much plant and animal life from the canal-like stretch of a river.

Furthermore canals link up all the main river systems draining central England. Boycott (1936) writes: 'And about the middle of last century a snail could start in the Thames at London and travel in uninterrupted water to Norfolk or Leeds or Kendal or Newtown in Montgomery or Hereford or Trowbridge, or by slipping into the upper waters of the Avon in the Vale of Pusey even to Christchurch or Southampton.'

6. Animals and Plants

The environmental background having been sketched, it is necessary to devote this chapter to an account of the different kinds of animals and plants which occur in fresh water.

We start with four paragraphs written for those naturalists who have not had a biological training, and who have studied only vertebrates. The number of different kinds – or species to use the proper term – is usually much greater in an invertebrate than in a vertebrate group and, since the animals are so much smaller, the differences are less obvious, and sometimes undetectable by the naked eye. Many people are surprised to learn that the word mosquito, for example, covers in Britain no less than forty different species. We have endeavoured to give some idea of the scope of the freshwater fauna by mentioning the number of known species in each group.

Latin names are unavoidable, since only a few species have an English name. Invertebrates are usually referred to by two names, and the first of these, the genus or generic name, corresponds to the surnames in the human community; though the Smith difficulty has been avoided by making a rule that the name of every genus must be different. The second name is the species name, and, like our Christian names, many are used over and over again; for example, *lacustris*, which means pertaining to a lake, will be encountered often in the following pages applied both to animals and plants. Sometimes a third name denotes a small, usually geographical, variation.

Anything of which there are many kinds must be grouped into higher categories for the sake of convenience; pence grouped into the higher category of pounds is a familiar example. Animal species are grouped into genera, genera into families, and so on through orders, classes, and phyla, with sub-divisions interpolated if necessary. The classification is based on structure and anatomy. The names of all these groups have a Latin form but, particularly in a work of this sort, they are frequently anglicized. Thus 'Insects' is more usual than 'Insecta', or 'Vertebrates' than 'Vertebrata'; and the midge family, for example, is referred to in the following pages as the Chironomidae or the Chironomids indifferently.

One of the main difficulties of a book of this nature is not so much to impart what is known, as to leave the reader with a fair impression of what is not known. Students of vertebrates or of one of the few well-known invertebrate groups, e.g. the Lepidoptera, may be surprised to find that, of the 61 species of water-bug in Britain, no less than 6, or 10%, were discovered in the decade before the war. The water-bugs may be taken as an average group, not as well-worked as some nor as neglected as others.

Most of the phyla are represented in fresh water, the total number of species is great, and it is not possible in these pages to do much more than indicate which of the main groups occur. As each group is mentioned, it is followed by figures in brackets. The first, sometimes the only, figure is the total number of species known from the British Isles, and the second the number which has been recorded from Windermere and the tarns and streams around it. The total for the British Isles is taken from the recent check-list of British insects by Kloet and Hincks (1945) and, for other groups, from the most recent available check-list or monograph. Some of the figures lay no claim to exactness; their purpose is to give an idea of the size of the group. The totals for Windermere district are taken from the files of the Freshwater Biological Association, and are given only for groups which have been studied fairly intensively.

It is convenient to begin this summary account of the freshwater fauna with the most advanced phylum, the Chordata, which includes all those animals with a backbone. In it are the fishes (45 British species, 15 recorded from the Lake District), and various amphibians, birds, and mammals which, although not fully aquatic, live in close association with water, and are noticed in Chapter 10.

The phylum Arthropoda derives its name from the fact that its members have an external skeleton and jointed limbs, the lobster being an example. A large number of aquatic animals belong to this phylum. Of the classes in it, the first, the class Crustacea, includes the Cladocera (90 British and 59 from Lake District) or water-fleas, and the Copepods (105) two groups of small animals. The plankton, that is the inhabitants of the open water of lakes and ponds, consists largely of members of these two groups, though some of them dwell among weeds. The creature known as the fish-louse (*Argulus*) is a degenerate Copepod. Another group is the Ostracoda (88 and 25), small animals to be found mainly among plants or in vegetable debris. At first sight it appears that the small, kidney-shaped seed of some plant has been taken, but, if left undisturbed, the object is discovered to consist of two valves, which open sufficiently to permit the extrusion of limbs with which the creature propels itself. Larger Crustacea are *Gammarus* (4 and 1), the freshwater shrimp, and *Asellus* (5), the water hog-louse, both of which are common, and *Cheirocephalus* (1 and 0), the fairy shrimp, which is rare and usually to be found only in muddy pools which dry up. Larger still is *Astacus* (1 and 1), the crayfish, which is the largest British freshwater invertebrate.

The class Arachnida, comprising Arthropod animals with eight legs, includes the water-spider and the water-mites or Hydracarina (234), small, round, often highly coloured creatures to be found in any habitat in fresh water. Many of the adults have long hairs fringing the legs and are able to swim; the immature stages are parasitic on insects and may often be found on the adults of such creatures as mosquitos, having attached themselves while the perfect insect was emerging from the pupal skin.

The class Insecta is characterized by the possession of six legs, and usually one or two pairs of wings as well, and it must be noticed in more detail as so many freshwater animals belong to it. The young stages in some orders have the same general build as the adults, and differ from them only in their smaller size, and lack of complete wings and genital organs; they are known as nymphs. The stoneflies, order Plecoptera (31 and 24), the mayflies, order Ephemeroptera (46 and 25) and the dragonflies, order Odonata (43 and 13) all have aquatic nymphs, but the adults live out of the water. The word 'mayfly' is employed, as it is commonly used by naturalists, to cover the whole order; fishermen restrict it to one stage of one genus (*Ephemera*), and leave the whole order and many of the species in it nameless. The bugs, order Hemiptera, are mostly terrestrial, but a certain number (61 and 40) live in or on the water throughout life, though they may occasionally take to the wing and fly from pond to pond.

The young stages of other orders of Insecta are quite unlike the adults and go by the name of larvae. A larva does not give rise directly to an adult as a nymph does, but enters a quiescent stage, the pupa (known as a chrysalis in butterflies) in which its tissues are broken down and reassembled into those of the adult. Orders with larvae are the caddis-flies, Trichoptera (188 and 108), all of which are aquatic in the larval, and terrestrial in the adult stage; the beetles Coleoptera (270 and 65), of which a few families are aquatic throughout life, like the bugs; and the flies, Diptera, of which several families have aquatic larvae. The family Culicidae, mosquitoes (37 and 7), is important because of its relation to disease, and the family Chironomidae, biting and non-biting midges (507), is important to students of lakes because its members are so widespread and numerous; lakes have been classified according to the kinds of Chironomid occurring in them. A few species in the orders Lepidoptera (butterflies and moths), Neuroptera, and Megaloptera, have aquatic larvae, and a small number of Hymenoptera (ants, bees, and wasps) parasitize aquatic larvae.

This ends the Arthropods and we come next to the phylum Mollusca; it has many marine representatives, but some, both 'snails' (43 and 16) and 'bivalves' (26 and 15) occur in fresh water. The Annelida, or true worms, are also marine, terrestrial, and freshwater (about 200), and the phylum includes the leeches (11 and 9). Other phyla are the Rotifera, wheel-animalcules (just over 500), all freshwater; the Platyhelminthes, mostly parasitic but with two groups, the little-known Rhabdocoeles (about 200) and the Planaria (8 and 6) occurring free in the water; and the Nematoda, round-worms, mostly parasitic, but with some members wholly freshwater and others with one stage in a complex life-history in fresh water. Finally the simplest animals of all fall into three phyla; the sponges, Porifera (5 and 2); the Coelenterata (6 and 3), which include the marine anemones, corals, and jellyfish, but which in fresh water are represented by simple polyps only, though there is one medusa (jellyfish form) which is found very rarely in British inland waters; and the Protozoa, or single-celled animals, which abound in fresh water. The Bacteria, assigned definitely neither to the plant nor to the animal kingdom, play an important, though as yet little understood role in the economy of all fresh waters. Most kinds are far too small to see except with the aid of a high-powered microscope, but some cluster in such vast numbers that the organisms themselves or their products become obvious. The sewage 'fungus', for example, is really one of the Bacteria, and the evil-looking reddish or yellowish substance often to be seen in springs and

1. Ford Wood Beck, a torrential headstream

2a. Water-bottle and Friedinger Winch. b. Reversing Thermometer and Friedinger Winch

3a. Saw-cylinder sampler. b. Jenkin surface-mud sampler

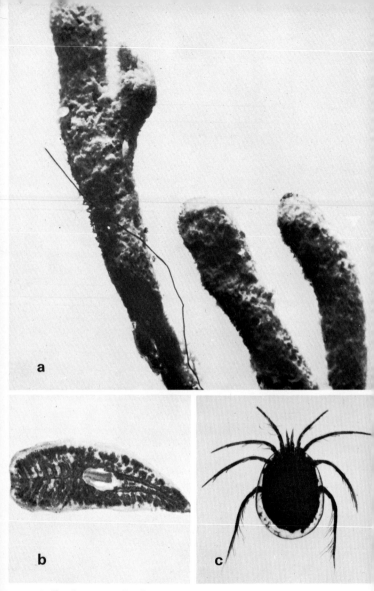

4. Freshwater animals
a. Freshwater sponge, *Spongilla lacustris*. *b.* Flatworm, *Dendrocoelum lacteum*. *c.* Mite, *Limnesia maculata*

5. Freshwater Crustacea
a. Freshwater shrimp, *Gammarus pulex*. *b.* Fish-louse, *Argulus coregoni*; male, dorsal aspect. *c.* Water hog-louse, *Asellus* sp.

6. Crayfish

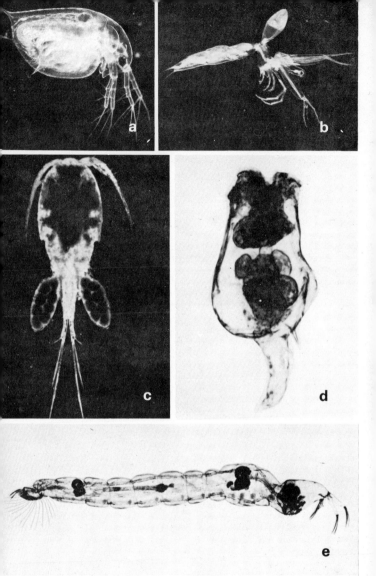

7. Plankton animals
a. Water-flea, *Daphnia magna*. *b*. *Leptodora kindti*. *c*. *Cyclops* sp.
d. Rotifer, *Brachionus calcyflorus*. *e*. Phantom larva, *Chaoborus* sp.

8. Freshwater Mollusca
a. Great pond snail, *Lymnaea stagnalis.* *b.* and *c.* River snail,
Viviparus viviparus. *d.* Swan mussel, *Anodonta cygnea*

seepages is a product of innumerable iron bacteria.

The main divisions of the plant kingdom are the flowering plants, the ferns, the mosses and liverworts, the fungi, and the algae. Animals living at the water's edge, either just in it or just not in it, can move up or down as the water level rises or falls. There is, therefore usually little difficulty in deciding whether they are aquatic or not. Plants cannot move, and there are many which grow at the water's edge in the zone which is sometimes flooded and sometimes dry; and so it is difficult to state exactly how many flowering plants come within the province of the freshwater biologist, but the number is probably not far short of 200.

Isoetes (the quillwort), *Pilularia* (the pillwort), *Azolla*, and *Equisetum fluviatile* (the horsetail), are the only aquatic representatives of the ferns (using that term in its widest sense). Mosses and liverworts often form extensive swards where the substratum is rocky or stony, and they profoundly influence the fauna by providing a foothold for animals which otherwise would be swept away by the current. Fungi are well represented, though by small and inconspicuous species only; they play an important part in decomposition, and some cause disease among higher forms. Included among aquatic fungi is a group called Chytridiales, the members of which are small and simple, but important in the general scheme because of their effect on the algae, which they parasitize. The least evolved group, the algae, lives mainly in the water, and in fresh water the green algae are rich both in species (over 2,000) and numbers, and of great importance as primary producers of living material. Some of the main groups are mentioned by name in subsequent pages, and require brief description here, though it is beyond the scope of this book to notice all of them and the differences by which they are distinguished. Diatoms are contained within a case of siliceous material, the flinty substance of which glass is made, and the case is in two halves which fit together like a box and its lid. There is often a delicate and beautiful sculpturing, and it was therefore the shell which caught the attention of the early microscopists, and led to the name. The individual is usually a single cell, but one of the commonest, *Asterionella*, is made up of a number of cells, loosely joined together at one end and radiating like the spokes of a wheel. Desmids are also usually single-celled, and each cell is commonly symmetrical about a central constriction, and of a characteristic and frequently beautiful shape. The individual blue-green alga is usually a number of cells in a common mucilage sheath. The colour of some of the species has a bluish tinge, and a discrete nucleus is absent. Some of the algae have small lashing filaments or flagella by means of which they progress, and these forms may be single, or united in spherical colonies. Certain of them occupy the border territory between the plant and animal kingdoms. Other algae consist of a row of cells joined together in a filament, and it is this type which will be most familiar to the observer who does not possess a microscope, since it produces the tress-like growth often to be seen on rocks and other objects in the water.

The scheme of classification which has been briefly outlined above is based on structure, and on the probable evolutionary relationships of organisms. It is fundamental, but, in a work concerned chiefly with the organism in relation to its environment, a grouping based on this relationship has a wider practical significance. There are five main groups in lakes and rivers, and it is not possible here to do much more than mention the main components of each; but it is hoped that the illustrations will give an outline picture of the groups discussed. The groups are:

> Surface forms,
> Stone- and weed-dwelling forms,
> Floating forms,
> Swimming forms, and
> Burrowing forms.

Of the Greek names which biologists apply to these groups the only one which cannot conveniently be ignored is 'plankton' for the floating forms.

The *surface dwellers* include the pond-skaters or water-striders, the water-measurer, some smaller bugs (Hemiptera) which have no English names, and the whirligig beetles. The pond-skaters or water-striders (*Gerris*) are rather narrow bugs with the middle and hind pairs of legs greatly elongated. These legs end in small pads which support the insect on the surface film, and the claws are displaced from the usual terminal position. The fore legs are used for seizing food. *Hydrometra*, the water-measurer, also known as the water-gnat, rather unfortunately since it has no relationship at all with the true gnats, is very slender, has a remarkably elongated head, and uses all six legs for walking. The piercing and sucking mouth parts are also used to secure the prey, and the extra movability which this entails is stated to be the reason for the elongated head.

The whirligig beetles (*Gyrinidae*) are small steely-black oval insects which are nearly always found in swarms, and which spend their time whirling rapidly around upon the surface of the water. The eyes are divided into an upper and lower portion, and the beetles are said to be able to see objects above and below the water surface simultaneously. The legs are shortened and flattened to serve as paddles. The larvae are truly aquatic. The adults live upon the surface but disappear into the water with surprising suddenness when alarmed.

Certain larvae of the Diptera, notably the mosquitoes, live in the water but attach themselves to the surface by a projection at the hind end of the body where the openings of the breathing tubes are situated. Some of them do not leave the surface unless they are scared, but others spend long periods feeding on the bottom of the pools which they inhabit.

The *stone- and weed-dwelling forms* are numerous and diverse. All the stonefly (Plecoptera) nymphs, of which the largest are known to fishermen as creepers, belong here. They are crawling forms, somewhat reminiscent of an earwig in general appearance though instead of forceps at the end of the body they have two long tapering 'tails' known as cerci. Most are found under stones in streams or along exposed lake shores, and only a few small species are found in typical pond conditions with a thick growth of rooted plants. The mayfly nymphs (Ephemeroptera) are similar but more diverse in structure. They have paired flaps down each side of the body, commonly known as gills, though, as will be shown later, as an indication of function this is a misnomer, and they may be distinguished at once by their three 'tails'. Some of the stone-dwelling species are flat, well adapted to cling to a stone in a current, but ill-equipped to swim; the typical weed-genus, *Cloëon*, on the other hand, is a good swimmer and propels itself by undulations of the body made more effective by the row of close-set hairs down each side of each tail.

All the dragon-fly nymphs belong to this community. The appendages at the hind end of their bodies also number three but are otherwise quite unlike the cerci of the mayflies. In the small dragonfly nymphs they are flattened and commonly, though again erroneously, known as gills, and in the large dragon-fly nymphs they are no more than triangular prolongations of the end of the body. The most characteristic feature of all dragon-fly nymphs is, however, the 'mask' or hinged labium, which can be extended suddenly in front of the head to seize some prey, which, unsuspecting, passes within reach of the hiding-place where the nymph is lurking. The larger ones can, on occasion, make a short but rapid dart in pursuit of some quarry just out of reach by suddenly contracting an enlarged chamber near the end of the alimentary canal. The smaller forms swim rather ineffectively by undulating the body from side to side.

Caddis larvae (Trichoptera) can usually be distinguished at a glance because all the body, except the head and the part bearing the legs, is protected by a case made of vegetable fragments, stones, grains of sand, or other suitable material, and the whole body is withdrawn inside the case if danger threatens. At the hind end of the larva there are two appendages known as false legs, and it is by means of these that the insect keeps hold of its case. These organs are characteristic of the whole order although some larvae do not have a case at all, but live free among stones or weeds, where most of them spin a net to catch their prey. Most of the case-bearing caddisworms can do no more than drag their cumbersome houses behind them, but some (*Triaenodes*) make a neat slender case of pieces of leaf stuck together into a flat-surfaced cone. Their hind legs are elongated and fringed with hairs so that they can be used as swimming organs.

The animals so far mentioned obtain oxygen from the water. Most have extrusions of the body wall of one kind or another, and these structures have been known for many years as gills. There is good evidence that many play no special role in the absorption of oxygen from the water and their real function is not known; though the 'gills' of certain mayflies keep a current of water flowing over the surface of the body and, since this is the seat of oxygen absorption, a current over it is valuable when the concentration of oxygen in the water is low. The so-called gills of mosquito larvae, which have a respiratory system open to the atmosphere, play a part in the maintenance in the body fluids of a higher concentration of salts than exists in the water inhabited by the animals. The bugs and the beetles are dependent on atmospheric oxygen and must swim up to the surface periodically to renew the bubble of air which they always carry about with them. The hind legs of both are elongated, flattened, and fringed with close-set hairs, modifications which have converted what was once a walking organ into a highly efficient paddle. The very youngest stages of the bugs depend on oxygen in solution but otherwise the immature forms of both groups require oxygen in gaseous form like the adults. The young bug is like the fully grown bug in general shape and build, but the young beetle is totally different. The familiar larva of *Dytiscus* is a spindle-shaped creature with six well-developed legs, two small appendages at the hind end of the body where the breathing tubes open, and a pair of large sickle-shaped jaws, possession of which immediately distinguishes it from any other larva. Other species have different forms and no single obvious feature is characteristic of the whole group.

Some other genera of beetles, the best known of which is *Donacia*, are terrestrial as adults but have larvae which dwell in the water. They are quite different from those just described, being grub-like and inactive. These larvae cannot utilize oxygen dissolved in water but they do not require to come to the surface for air, as they possess a modification whereby they can tap the air-containing tubes in the roots of plants. The same habit is found in a genus of mosquitoes and in two other families of flies.

Dipterous larvae contribute prominently to the stone- and weed-dwelling component of the fauna. They are diverse in form, but one common feature is the absence of jointed legs such as all other larvae possess. The Chironomid family is almost ubiquitous in fresh water, but, as it is more typical of the burrowing fauna, description of it is deferred. Other dipterous larvae are grub-like creatures which may belong to the daddy-long-legs (*Tipulidae*), horsefly (*Tabanidae*), or one or two other families. There is an interesting series in the first, from the larva which dwells in the comparatively dry soil of lawns (where incidentally it is a notorious pest on account of its habit of eating grass roots) through one which dwells in the damp mud of marshes, to species which are truly aquatic. The larvae have occupied these quite different types of habitat without any great modification of structure.

An interesting adaptation is seen in the rat-tailed maggot which, when full-grown, becomes one of the hover-flies (*Eristalis*). The body is grub-like but provided on the underside with pad-like projections known as false legs or pro-legs. With these it walks about on the bottom of highly polluted localities, obtaining oxygen meanwhile through a remarkable long telescopic tail at the end of which the breathing tubes open.

In streams and rivers, particularly where the current is strong, larvae of *Simulium*, the buffalo gnat, are often abundant. The larvae have a characteristic dumb-bell-like shape and there is a pro-leg just beneath the head. This pro-leg and also a big round pad on the bottom of the animal are provided with hooks. The larva spins a web over the surface of some rock or stone, and then fixes itself by means of its basal attachment organ, allowing the head to trail free in the water. The mouthparts are provided with a rake-like arrangement of bristles by means of which minute particles carried down by the current can be secured. When the animal desires to move, it attaches itself by the pro-legs and swings the body to a new position, and so, holding fast by the front and back alternately, it proceeds rather like a 'looper' caterpillar or some of the leeches.

These are the more important insects dwelling among the weeds or on stones. Many other invertebrates inhabit this part of the freshwater environment. *Gammarus*, the freshwater shrimp, is very widespread and often numerous. It is flattened from side to side and arched from head to tail. It can swim quite effectively, but spends most of its time in the debris on the bottom. Characteristic is the male habit of carrying a female around with him almost permanently. The sand-hopper, found beneath piles of sea-weed on the seashore, is a relative of *Gammarus* likely to be familiar to many. *Asellus*, the water hog-louse is also widespread. It is a more sedate creature, and a not very rapid walk is the only form of locomotion of which it is capable. It is clearly related to the 'slaters' or wood-lice of the land, but is more slender. Some smaller Crustacea also belong to this community but, except for the Ostracods, they are more typical of the plankton, and are described there. Mites also inhabit this part of the freshwater environment, and there is scarcely any type of bottom which will not yield some of the many different species.

All the freshwater snails belong to this group. Though varying in size and shape, most of them are 'snail like' and require no description; but there are two species which have lost the coiled form and come to resemble the familiar limpets of the seashore both in aspect and way of life, though, like most freshwater animals, they are much smaller than their marine counterparts. Some snails (e.g. *Viviparus*) have gills which may or may not protrude from beneath the shell, and they have invaded fresh water from the sea. Many freshwater snails (e.g. *Lymnaea*) breathe by means of a lung, and are descended from land-living ancestors.

Most of the Annelida or true worms belong to the burrowing group, but all the members of one order, the Hirudinea or leeches, dwell among weeds or stones. A sucker at the head end and another at the tail end of the body is characteristic of the leeches. By using these suckers they can progress from place to place rapidly, 'looping' like a Geometrid caterpillar, but most can also swim quite effectively with an undulating motion. Some suck blood, but most swallow small invertebrates whole.

The flatworms or Planarians are flat but they are not like ordinary worms at all, since they are unsegmented. The body is long and narrow and seldom more than half an inch from the front of the head, which is usually rather square, to the tip of the tail, which is pointed. The creatures are coloured black, brown, or white, and they are familiar objects gliding about on almost any type of substratum except, perhaps, soft mud. There is something rather mysterious about their progress for the organs of locomotion are too small to be seen with the naked eye.

Sponges are not uncommon encrusting stones or branches submerged in the water. Close examination reveals that they are perforated with holes of various sizes and that they also resemble in texture the bathroom object, though much smaller. The Coelenterata are represented commonly in fresh water by *Hydra*, which is like a sea anemone in general build but much smaller, relatively more slender, and with longer tentacles. It is a frequent object hanging from submerged plants but, as it is quite small, no thicker than coarse thread when fully extended and rarely as much as an inch long, it is overlooked except by those who know exactly where to search for it.

Protozoa abound upon the surface of stones and weeds, but hitherto they have not received much attention.

The *plankton* is remarkable because it includes only a single insect, the 'phantom larva', *Chaoborus*. This larva, which gives rise to an adult closely resembling a mosquito, though it does not bite, has several peculiar features. The body is almost transparent and the only conspicuous parts of it are the eyes and two pairs of air-sacs, one in the thorax and the other near the end of the abdomen. These sacs can be expanded and contracted, apparently by some substance which is secreted into the body fluid as a result of nervous stimulation, and by means of them the animal can rise, sink, maintain its position in the face of changing atmospheric pressure, or remain level when the front end is weighed down by some organism which has been captured for food. The organs with which the prey is secured are the antennae or feelers, and this modification of these organs is unusual among insects.

All the other larger (this is a relative term, they are all small animals) members of the plankton belong to the Crustacea, and fall into two groups, the *Cladocera* and the *Copepoda*. The Cladocera are often called water-fleas because some kinds resemble real fleas in size and shape, and progress in a series of jerks; they do not, of course, bear the slightest true relationship to fleas. Both Cladocera and Copepoda have been called floating forms and distinguished from the swimming forms, though actually they are capable of a fair amount of locomotion. This, however, is mainly in an up-and-down direction and they are not capable of any considerable progress in the horizontal plane. The organs of locomotion in both groups are the antennae or 'feelers', but here resemblance ceases. The Cladocera have typically two 'shells', not comparable with the shells of a bivalve mollusc but not unlike them in the way they enclose the body, except for a narrow groove along the ventral surface of the animal. Within the shells a series of hairy limbs move constantly to and fro and maintain a current of water in which minute animals, plants, and particles of debris are conveyed to the mouth. Where the shells join along the back there is a chamber in which the embryos are kept. Some species have the shell much reduced and the antennae more developed (e.g. *Leptodora*), and are carnivorous, preying on other planktonic forms.

The Copepods are pear-shaped and without shells. Many of them are carnivorous. The eggs are attached to the hind end of the body, usually in two oval masses projecting on either side, though sometimes there is but one in the middle.

Also important in the plankton are the Rotifera or wheel animalcules. They are of rather simple structure and small or very small size. Their characteristic feature, giving origin to the name, is a crown of small hair-like excrescences called cilia, which by continual beating give the appearance of a rotating wheel and maintain a current of water which conveys minute particles of food to the mouth.

Protozoa, or single-celled animals, are abundant in the plankton and the number of different species is large. Some (Ciliata) have cilia like those of the Rotifers and may indeed resemble a rotifer closely; others have one or two longer loco-motory organs; these are known as flagella, and the group as the Flagellata. Finally there are the Heliozoa or sun-animal-cules, which have long immobile threads of protoplasm radiating in all directions.

The *swimmers* in fresh water include only the fish and the higher vertebrates, amphibia, birds, and mammals, which need no description here, and so we reach the last group, the *bur-rowers*. In this group insects are represented by a great many species of the fly (Diptera) family Chironomidae. The adults are midges but most of them do not bite and are larger than the small black fly which is such a pest in the evening in some parts of the country. Superficially these larger ones are not unlike mosquitoes, but, whereas it is the hind legs of mosquitoes, it is the front legs of Chironomidae which are longer than the other pairs. The larvae are elongated, somewhat worm-like creatures with a pair of unjointed legs just behind the head and another at the end of the body. The head is of a characteristic shape and there is a variable number of sausage-like protuber-ances, known as gills, on the last or the last two segments. Some Chironomid larvae contain haemoglobin, the red pig-ment of vertebrate blood, and are known as bloodworms. Most species live in tubes made from the mud which surrounds them, but all can swim, although not very effectively, by lash-ing movements which throw the body into a figure of eight. This method of progression is diagnostic of the family.

The larvae of the alder-fly, *Sialis*, and the nymphs of the mayfly, *Ephemera*, dwell in burrows in mud or fine sand. The two are not dissimilar superficially but *Ephemera* has the three tails characteristic of its order, whereas *Sialis* has but one. *Ephemera* has feathery 'gills' curved over its back and *Sialis* has simple filamentous 'gills' projecting from the sides. When teased, *Sialis* opens its jaws, elevates its tail and goes back-wards, a syndrome which is characteristic. The powerful jaws are a clear indication of the carnivorous propensities of this larva. Pupation is believed to take place always on land, and, if this is invariably true, some larvae in Windermere must migrate two hundred yards or so to find suitable conditions.

The Annelid worms are important among the burrowers. Some are not unlike the familiar earth-worm, but most are smaller and more thread-like. Finally there are the bivalve mol-luscs, ranging from the swan-mussel, which may be as much as seven inches long, down to the pea-mussels (*Pisidium* and

Sphaerium), some of which are only about the size of a capital 'O' on this page.

Protozoa and other microscopic animals are probably important but at present neglected members of the mud fauna.

These are the five groups into which fresh-water animals may be classified on habit, and their most important members. Distinction between them is, on the whole, clear-cut. Some of the surface dwellers may occasionally dive beneath the surface, and sometimes a snail or a flatworm may be seen crawling along the underside of the water-surface, but there is no doubt about the group to which these animals really belong. Some of the stone- and weed-dwellers are good swimmers but they spend most of their time attached and are never found out in the open water far away from their substratum. Cladocera and Copepoda are found both in weeds and far out where they must spend all their lives clear of anything to rest on, but, in general, the species found in the two types of habitat are not the same.

Certain stone-dwellers provide the only instance where there may be some doubt. For instance, the nymph of the large black and yellow dragon-fly, *Cordulegaster boltonii*, is found buried in gravel, and nymphs of the genus *Agrion* may lie buried in mud. But neither make any sort of real burrow, and in feeding habits both belong clearly to the stone- and weed-dwellers, so it is to this group that they have been assigned here.

The plants may be treated in a similar, though not identical manner, and divided into surface dwellers, species attached to the substratum, and plankton. Though some of those which possess cilia or flagella may be seen passing across the field of a microscope at what appears to be high speed, none can make such progress in a lake that it is not at the mercy of the slightest current, and it may be said that there are no true swimmers in the vegetable kingdom. Nor does any plant qualify as a burrower though many have a root-system embedded in the substratum. The surface dwellers float at the surface with no attachment to the bottom of the pond or lake. The total number of species is small but they are drawn from widely different parts of the plant kingdom. There are two liverworts, *Riccia* and *Ricciocarpus*, the water 'fern', *Azolla*, and, among flowering plants, the frogbit, *Hydrocharis morsus-ranae*, and five species of duckweed, *Lemna*.

Almost any substratum, provided it is sufficiently well illuminated, is colonized by vegetation. Rock faces, boulders, and large stones are covered with algae which may trail in the water in hair-like tresses or form a thin coating of slime. Several species of moss and liverwort also attach themselves in such places, and in a stony river, provided the stones are large and not too liable to shift in a flood, there may be extensive and dense beds formed by *Eurhynchium rusciforme, Fontinalis* spp. and others. Sandy and muddy bottoms are colonized predominantly by higher plants, though mosses and algae may occur, and the peculiar stoneworts, *Chara* and *Nitella*, which are algae though their form resembles superficially that of more advanced plants, may cover wide areas. It is convenient to divide the rooted plants into those which are submerged, those which have floating leaves like the water-lilies, and those which are emergent, that is the reeds, rushes, and sedges. There are some plants which have both submerged and floating leaves, or both floating and emergent leaves, but on the whole this classification is a useful one. It is usual to find the emergents in the shallowest water, the submerged plants in the deepest water, and the floating-leaved plants in a zone between. The floating-leaved species must not of course be confused with the floating species, or surface-dwellers, mentioned on the previous page.

The planktonic forms of plant life are numerous and diverse but all belong to the Algae.

7. The Organism in its Environment

The purpose of Chapter 6 was to give a picture showing the different kinds of organisms to be found in fresh water, setting them out first according to the classification based on structural affinity, and then according to a grouping based on their relationship to the environment. The present and subsequent chapters take different kinds of freshwater biotopes and describe the animal and plant communities which dwell in each. The communities are treated very generally, because a work of this sort is not the proper place to give long lists of species. The authors take this opportunity to stress that the number of species in any one uniform part of the environment (here referred to as a biotope) is of a length that surprises the layman and often the biologist as well. It can easily fill several pages, and the long list of names conveys little to any save those closely acquainted with the biotope or with others very like it. Any discussion, even that by the original author, must be based on a limited number of species selected for one reason or another.

Windermere, being the best known biologically, is a convenient standard with which other lakes may be compared. It is, to recapitulate some of its salient features, 10½ miles (17 km.) long and about half a mile (800 m.) in breadth, so that, although the biggest in England, it is not a large lake as lakes go; but it is of such a size that the wind can raise waves of sufficient magnitude to erode any shore on which they beat. The greatest depth is 219 feet (67 m.) and the average depth is 73 feet (22·3 m.). Important data for the biologist are that the greatest depth to which sufficient light for rooted-plant growth penetrates is about 16 feet (5 m.), and 25 % of the lake bottom is covered by less than this depth of water; and that, if 32 feet (10 m.) be taken as the approximate average depth of the thermocline, then the volume of water below it, that is the hypolimnion, is twice the volume of water above it, that is the epilimnion. All the rocks in the drainage basin are relatively hard and deficient in lime, but most of the land immediately surrounding the lake is under cultivation, and there are fairly large human settlements at Ambleside, near the head of the lake, and at Windermere Town near the centre of the lake. The result is that, though the waters of Windermere are soft, the amount of nutrient salts is fairly high, and the lake is moderately productive. But, owing to the large size of the hypolimnion relative to the epilimnion, the oxygen in it never becomes depleted or even approaches depletion, and Windermere is morphometrically oligotrophic.

There are no surface plants in Windermere and the population of surface animals is poor. In extremely sheltered places, such as boat-houses, one of the larger species of pond-skater, *Gerris najas*, occurs in congregations, and there are also colonies of the whirligig beetle, *Gyrinus*. Some of the smaller water-bugs occur in the reed-beds.

As long ago as 1909 W. and G. S. West published lists of the minute floating algae or phytoplankton. They recorded 188 species in 11 lakes, drew attention to differences between the lakes, and referred to the periodicity of algae in Windermere, the only lake from which they had samples taken throughout a season. This work was continued by W. H. Pearsall and his father, who eventually recorded 205 species in 11 lakes. They showed that the number of specimens to be found during the course of the year is comparatively small in an unproductive lake and much larger in a productive one, and also that there is a big difference in the species found. The dominant species in unproductive lakes such as Ennerdale and Wastwater are desmids. In more productive lakes *Dinobryon* is important as well. In the most productive lakes diatoms are dominant, and in Esthwaite blue-green algae make a large contribution to the total too.

Asterionella is a diatom which reaches great abundance each spring in Windermere and Esthwaite, but which does not occur in Buttermere. It will not grow in water from Buttermere. However, if soil is added to this water, growth does take place. The nature of this substance which *Asterionella* must evidently have, and which occurs in many soils, remains unknown.

After many years of theory and speculation, the swift increase in the numbers of *Asterionella* and their equally rapid decrease were explained by Dr. J. W. G. Lund. The alga is present all through the winter, but numbers remain low because physical factors are adverse. In particular the days are short and, when the sun does shine, it never rises high above the horizon and much of the light is reflected off the surface. Moreover, because the lake is unstratified, an alga at the surface is liable to be swept down to a depth at which there is no light at any time of day. The low temperature, and the flushing effect of rain, greater than in summer because evaporation is less, are other unfavourable factors. There comes a time in spring when conditions are right for rapid multiplication and the algal population becomes very numerous. Then reproduction stops, algae start to die, and the numbers fall. Lund studied this rise and fall for many years, and observed that whatever else might be different, the concentration of silica in the water had always fallen to 0·5 p.p.m. at the time of the algal maximum. Figure 13, illustrating this point, is one of many from the works of Lund.

After the outburst of *Asterionella*, other species, presumably with different requirements, wax and wane in the same way, though the total number of algae present is never as great as at the time of the *Asterionella* maximum.

One year *Asterionella* suffered three setbacks. An unusual abundance of a blue-green alga slowed down its rate of reproduction, probably by shading it from the light. Numbers were washed out of the lake by an untimely flood. Infestation by a parasitic fungus was unusually heavy. The result was that two rival diatoms, *Tabellaria* and *Fragilaria*, used most of the silicate, and the maximum reached by *Asterionella* was much lower than usual. Lund offers an explanation of why *Asterionella* nearly always beats these two in what appears to be a race for the available nutrients. *Tabellaria* reproduces less frequently than *Asterionella*, which soon outstrips it. *Fragilaria*, on the other hand, has a similar rate of reproduction but, for some unknown reason, is scarcer in winter and accordingly starts the race with smaller numbers.

In other lakes it has been shown that other substances are in shortest supply and in consequence are the agents which bring algal reproduction to a halt, and the same is probably true in Windermere of species that have not been studied. Phosphate is an important nutrient that is often present in low concentration, but Mackereth has shown that *Asterionella* is able to store this substance. It takes up such large amounts in winter, when there is no shortage, that the supply, divided equally when a cell divides into two, is sufficient to last until the silicate runs out.

Some algae secrete substances that inhibit the growth of other species or even kill them. Workers at Windermere have not found any evidence that this occurs in the Lake District

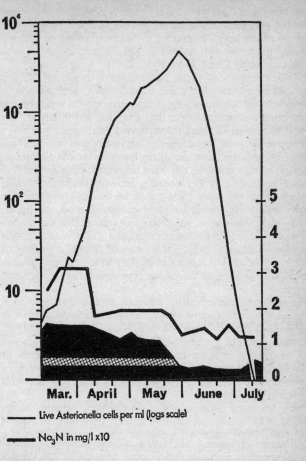

Fig 13. Numbers of *Asterionella*, concentrations of silicate and nitrate in Windermere North Basin in 1945 (J. W. G. Lund).

lakes, but Lund does not rule it out as a possibility elsewhere. The blue-green algae are noted for the production of toxins, some of which may be noxious to beasts drinking or humans bathing. Examples are quoted in the proceedings of a symposium edited by Dr D. F. Jackson (1964).

The best known of these 'anti-biotics' is penicillin. This is produced by the familiar green mould which is sometimes, to the vexation of the housewife, found growing on a pot of jam or an orange which has been left too long in the larder. The organism, known as *Penicillium*, having established itself on a nutritive substratum, produces this anti-biotic and prevents bacteria or other fungi from exploiting the same source of food.

A remarkable instance has been described quite recently. A small alga known as *Prymnesium parvum* suddenly became very abundant in some carp ponds in Israel. The carp died rapidly, killed apparently by some substance produced by the *Prymnesium*. The sequel was even more remarkable, for a comparatively short and simple investigation discovered a simple and effective remedy. Nitrogen is essential for the organism, as it is for any other, but it must be in the form of nitrate; ammonium salts in low concentration kill *Prymnesium* in a short space of time.

The difficulties confronting animals that live in the open water, and the adaptations by which they have been overcome, were noticed briefly on p. 22. Mr W. J. Smyly has studied the zooplankton of the English Lakes, and the result of his and other researches is a list of some 8 species of Cladocera or water fleas, 5 species of Copepoda, 34 species of Rotifera or wheel-animalcules, and 13 species of Protozoa, together with the insect *Chaoborus*.

There are some peculiarities of distribution, notably the restriction of *Limnocalanus* to Ennerdale (p. 147) but in general it is not possible to make out a change in the species list as one passes from unproductive to productive lakes, a change exhibited by the communities of almost every other part of the lake and by the phytoplankton. The number of individuals, of course, increases along the series.

Some species in all the groups are carnivorous. Among the rest it is unexpected to find that the largest animals eat the smallest plants, and the smallest animals the largest plants. The Cladocera feed by straining the water by means of a mechanism that rejects all but the very minute algae. They probably consume many bacteria and tiny fragments of organic matter, and possibly even absorb organic matter in solution. One species of rotifer is able to tackle algae as large as *Asterionella* and recently Dr H. M. Canter (Mrs J. W. G. Lund) has shown that certain Protozoa can ingest algae of this size. This work is on the threshold, and to Dr Canter belongs the distinction of pioneering not only this line, but the study of the minute fungi that parasitize algae. Generally neither rotifers nor Protozoa make serious inroads into populations of algae, and the crop is largely unexploited. Dead algae decompose either in the water or in the mud, or are washed out of the lake.

Some species of water-flea reproduce asexually throughout the summer, and the offspring from this method of procreation are all females. In the autumn males are also brought forth, and then sexual reproduction ensues. The fertilized eggs have a hard resistant coat and lie dormant for a long period before undergoing further development. Of two rival schools of thought, one maintains that the appearance of males and sexual reproduction is a rhythmical phenomenon which takes place after a certain amount of asexual reproduction, and the other that it is a definite response to the onset of unfavourable conditions. Work by Kaj Berg and C. H. Mortimer supports the latter view. In cultures of various species asexual reproduction continues as long as conditions are optimal, which may be much longer than ever happens in the field. Unfavourable conditions in which males are produced are a lower temperature, reduced food-supply, and overcrowding.

Some Cladocera of the water-flea type have in winter a round front end and a relatively short 'tail' projection, but in summer a pointed front end and a longer 'tail'. The small round *Bosmina* may have in summer a hump which is lacking in winter, and a longer proboscis. Examples of these differences are shown in Figure 14, which is taken from Wesenberg-Lund's (1908) big paper on the Danish plankton; but it must be mentioned that, for the sake of illustration, extremes have been chosen from among his very numerous drawings. Some of the algae also show a greater development of spines and other projections from the body surface in specimens taken in summer. All these summer forms have a greater surface area and therefore sink more slowly, since they offer more resistance to the water; all plankton forms, except a few which have bubbles of gas in them and the insect, *Chaoborus*, which has hydrostatic organs, are slightly heavier than water and tend to

current of the epilimnion, whereas the plain forms are carried up to the surface and eaten in greater numbers by fish because they are more easily seen in the brighter light. It is difficult to observe in nature a small animal which lives in a big volume and there are theoretical objections to this suggestion. It demands an epilimnion circulating at such a speed that cladocerans could keep station swimming against it and with an absence of turbulence that must be rare. J. L. Brooks believes that, as the helmet is more transparent than the rest of the body, it enables the animal to grow to the same size but present a smaller target to fish. It is not clear why, if smallness is advantageous, there should not be a straightforward reduction in size without any outgrowths. A complete explanation is probably not far off, but at the moment, nobody has succeeded in tying up all the loose ends.

One of the most remarkable activities of the Crustacean members of the plankton is their constant movement up and down, a phenomenon which has been observed and investigated in many parts of the world. At midday the animals are in greater abundance at some distance from the surface, and in the upper layers of the lake they are few. As the day draws on there is an upward migration, and round about midnight it is the surface layers of the lake which are most thickly populated. Then the return migration to the depths begins as night gives place to day (Fig. 15). The movements of the plankton

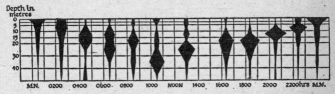

Fig. 15 The daily migration of *Cyclops strenuus* in Windermere. In the construction of the figures the catch at each depth has been converted to percentage of the total catch (After Ullyott, 1939)

and the movements of the sun are obviously related in some way, though whether the plankton organisms move to keep themselves in a zone where light of a certain intensity is always the same, or whether they react to some other stimulus as well is still a debated point. It has, for example, been suggested that the animals tend to move against the pull of gravity but avoid illumination above a certain brightness. Ullyot (1939) points out certain objections to this interpretation and has drawn attention to a close correlation, illustrated in Figure 16, between the depth at which the plankton is thickest and the depth to which rays of light at the blue end of the spectrum penetrate with a certain intensity. This work suggests that the movements are a reaction to light alone.

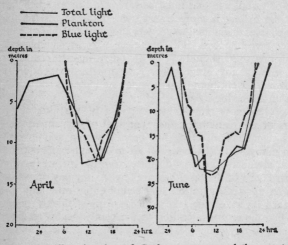

Fig. 16 Daily migration of *Cyclops strenuus* and the penetration of light in Windermere. The thick continuous line is obtained by plotting at hourly intervals the depth at which most specimens were caught (Fig. 15); the thin continuous line shows at hourly intervals the depth to which total light of an intensity of 32,800 erg/cm²/sec. in April and 108,000 erg/cm²/sec. in June penetrated; the thick broken line shows blue light treated in the same way, the intensity taken being 9,600 erg/cm²/sec. in April and 305 erg/cm²/sec. in June (After Ullyott, 1939)

It is difficult to believe that this travelling up and down, which must use up a large amount of energy, is entirely functionless, though at present it is not possible to do more than speculate about what end it may serve. It has been pointed out that the water in lakes is always streaming in a horizontal direction, and so any animal which tended to stay at one level would get carried to the side of the lake. By constantly moving up and down the animals will pass from a current in one direction at one depth to its counter-current in the opposite direction at another depth, and the effects of the various currents will cancel each other. But the phytoplankton does not undergo vertical migrations each day like the zooplankton, so the problem immediately arises of how the phytoplankton meets the same contingency. Most of the algae sink steadily and some observations made in Windermere one year showed that under ice the upper layers of the water became quite devoid of algae. It must be assumed that under usual conditions swirls and eddies set up by various causes keep the upper layers of the water continually supplied with algae.

The late M. M. Kozhov suggested that the zooplankton comes to the upper layers to feed by night and retires to darker layers by day to avoid being fed on.

To resume the account of life in Windermere it is logical to pass on from the plankton to the animals which feed on it. It is the main source of food for only one, apart from the carnivorous members of the plankton itself, and that is the char (*Salvelinus alpinus willughbii*). This fish inhabits the deeper parts of the lake, but some specimens come into the shallow water before Christmas to spawn. Others, belonging to a different race, lay their eggs in deeper waters in the early months of the year.

Also linked in the food-chain with the two preceding groups are the burrowers. In Windermere, Humphries records all the main types already enumerated with the exception of the large mussels. The nymphs of *Ephemera*, the Mayfly, represented by a single species, occur only where sand floors the lake, and have not been found at depths over 30 feet (9 m.). *Sialis*, the Alder-fly, also represented by one species, is found in mud, but it, too, is limited in depth and does not occur beyond 40 feet (12 m.). At this depth it may have to travel a considerable distance to reach dry land and, since it is believed always to pupate on dry land, it would be interesting to know more about this journey. At greater depths the fauna consists only of 8 species of midge larvae (Chironomidae), 2 species of true worms (Oligochaeta), 5 species of pea-mussels (*Pisidium*), and an occasional mite. These figures are based on revised lists, except that relating to Chironomids, which is based on larval identifications and may, therefore, be too low. *Chaoborus*, the phantom larva, is also present.

The creatures found in the deepest part of the lake live perpetually in an environment which, though not comparable with the depths of the ocean, is none the less remarkable for its

lack of variety; water movement is no more than an impercep-
tible creep, midday is scarcely less dark than midnight, and the
temperature range throughout the year may be no more than
from 4° C. to 6° C. (Fig. 1, p. 36). The inhabitants of
Windermere have not to contend with a lack of oxygen in the
summer, a physiological problem which dwellers in richer
lakes have to solve. The Chironomids do not spend all their
lives in the mud, but must make the long journey to the surface so that the adult may emerge and reproduce; it is a
journey fraught with peril, for the ascending pupae are preyed
on extensively by fish.

Finally there is the shallow region around the margin of the
lake. Conditions are more diverse here than in any other part,
and the fauna shows a corresponding richness in number of
species and variety of form. In order to avoid a controversy
which generates much heat among continental workers, who
disagree about the limits of the littoral and sublittoral zones in
relation to the water level, it is proposed here to refer to this
part of the lake as the zone of attached vegetation. The depth
to which it extends depends on the depth to which light of a
certain intensity can penetrate, and in Windermere this is
about 16 feet (5 m.).

Most of the shallow-water region of Windermere is so
exposed to wave action that the soil which rooted plants require is washed away and the substratum consists of stones, or,
in places, bare rock. On this there is a covering of algae.
Enormous lists of algae have been compiled, but work has not
yet reached a stage at which a coherent story of what the
important species are under any given circumstances can be
presented. Where there is some shelter, as for example in High
Wray Bay (Fig. 4, p. 44), *Littorella uniflora*, the lake wort, may
form a compact sward.

In more protected bays beds of emergent vegetation occur.
The first species to supplant *Littorella* is *Phragmites communis*, the common reed, and once it has become established it
exerts a profound influence on subsequent developments. The
changes which take place have not actually been observed in
any one reed-bed, but there is good reason to believe that the
more sheltered a bay the more rapidly will the vegetation in it
develop or evolve; and therefore from a study of the present-
day flora of a series of reed-beds, each more protected from
wave action than the last, the plant succession can be ascertained.

A thick reed-bed affects the environment in two ways; first, by dying down at the end of each season it covers the bottom with organic matter, and secondly, it impedes the flow of water. Vegetable debris decomposes very slowly unless it is well mixed with silt. The details of the chemistry of the process are not known but there is evidence that one important element which comes in with silt and speeds up decomposition is calcium. A reed-bed which colonizes a sandy bottom will, therefore, by dying down each year and impeding silt-bearing currents of water, gradually convert the substratum from one which contains little to one which contains much organic matter. Such bottoms are unfavourable to many species of plant, among them *Phragmites* itself, and the reed is gradually replaced by sedge, *Carex rostrata*. Bogmoss, *Sphagnum*, will enter the plant association later on and, as the continued accumulation of organic matter raises the soil, bog and then moorland plant communities will appear.

But it may happen that a reed-bed develops at the mouth of a river and silt is deposited upon each season's plant remains. The result is a greater degree of decomposition of the organic matter and the formation of a rich black mud. *Phragmites* can grow on this substratum and it will persist until, during spells of fine weather, it is growing on dry land. Sallows and other bushes join the plant community at this stage, and a fenland type of vegetation is produced. There is one such fen by Windermere where the stream draining Blelham Lake flows in, but the best known example is the fen at the head of Esthwaite Water, a locality which Professor Pearsall surveyed just before the First World War and at intervals of about fifteen years thereafter. His maps of the vegetation may be seen in Tansley's book (1939, pp. 640 and 641).

We turn next to the changes that may be observed at right-angles to the shore. The waves carry away the finer particles from the shallow water, and drop them in deeper water, the coarsest coming down first. The stones are therefore covered by gravel, which, in its turn, as the bottom descends, gives place to sand and then silt of ever-increasing fineness. Pearsall showed that, although light does play some part, the main factor that brings about a zonation of plants parallel with the shore, is the nature of the substratum. *Littorella*, which reproduces extensively by stolons, can colonize the area where the intensity of wave action is beginning to diminish. Below it on sand there is a zone of *Isoetes*, the quillwort, and below that on the finer soil, one of *Nitella*, a stonewort. On the fine silts are encountered great forests of the big *Potamogetons*, such as *P. perfoliatus*, *P. praelongus*, and *P. pusillus*.

In the unproductive lakes there is little fine silt and in consequence not much submerged vegetation except *Isoetes* and *Nitella*. In more productive lakes the percentage of *Potamogetons* increases. In Esthwaite there is a copious supply of fine silt and, associated with this, is an association dominated by species that are much less common in the other lakes. Another result of the rich silting in Esthwaite is the replacement of *Phragmites* by the reedmace, *Typha*, at the mouth of the main inflow.

The first study of the animals of the shallow water of Windermere was made by Moon who investigated the stretch of shore already described. On the bare rock face he found but two species; one of them, *Ancylus fluviatilis*, the fresh-water limpet, is in fact like a miniature limpet (though not related to those of the seashore) and is adapted to colonize just such a habitat as a bare rock face. The other is the ubiquitous *Lymnaea peregra*, the wandering snail. Where there are cracks affording a slight degree of cover, the fauna is augmented by two species of caddis larvae.

On the Bannisdale shore there is a much greater degree of cover to be obtained underneath the slates, and the fauna is much richer in species. In addition to the two snails already encountered, occasional specimens of *Planorbis carinatus* (the keeled ramshorn), *P. contortus* (the twisted ramshorn) and *P. albus* (the white ramshorn) may be found. Flatworms, mites, leeches, *Gammarus* (the fresh-water shrimp), and *Asellus* (the water hog-louse) also occur in small numbers. The characteristic feature of the fauna is the variety of insect nymphs and larvae. There are 4 species of mayflies, 5 species of stoneflies, and about 18 species of caddisfly larvae; subsequent workers having bred out the adults, it is possible to give precise numbers for the first two groups, but the total of the last must remain approximate until the adults of the order have been studied. Chironomid larvae are also present in fair numbers. These creatures are almost ubiquitous in fresh water, but little is known about the species of the different biotopes on account of taxonomic difficulties. There are over 400 species in Britain; the adults can be named only by a specialist, and the larvae cannot be named at all in the present state of knowledge.

Some members of this fauna possess obvious structural

adaptations for life in this particular environment. The limpet shape of *Ancylus* has already been mentioned, and both *Ecdyonurus dispar* and *Heptagenia lateralis*, two of the Ephemeroptera, have flattened, spread-eagled bodies adapted for life on the surface of a stone. Other animals, as for example the stonefly, *Diura bicaudata*, do not occur in other biotopes although there are no obvious structural reasons why they should not do so. The restriction to one type of biotope, for some species at least, is attributable to physiological adaption or physiological restriction, a subject which is discussed more fully later, on p. 150. The rest of the fauna may be described as adaptable, rather than adapted, and it occurs in a wide range of habitats; *Lymnaea peregra* and *Gammarus pulex* are two extreme examples. Many of the Trichoptera (caddis flies) make their cases of stones, and such cases are more compact and heavier than those made of plant material, and therefore less liable to be moved about by wave action. It is, however, rash to leap to the conclusion that this represents adaptation; species which build stone cases may do so because they have formed the habit of utilizing the material most ready to hand in a stony environment. In the genus *Silo*, however, a row of larger stones is attached to the sides of the case, and this may perhaps serve to prevent the case being turned upside down.

A general feature of the fauna is a scarcity of swimming forms, particularly those which have to come to the surface for air, though a beetle, *Platambus maculatus*, is a characteristic member of the community.

The fauna of the Drift shore is similar, but it is augmented by some of the burrowers. The next habitat in the series, bare sand, is inhabited only by the specialized burrowing species – various Chironomid larvae, pea-mussels (*Pisidium*), worms, the nymphs of the mayfly (*Ephemera*), and the larvae of the alder-fly (*Sialis*); the total number of groups represented is small.

Colonization of sand by reeds produces sheltered conditions in which a more varied fauna is to be found. It includes many of the Bannisdale shore species, but there are some notable absentees, and some species not previously encountered. To take the absentees first: they are *Ancylus fluviatilis*, *Ecdyonurus*, and *Heptagenia*, the three animals with structural adaptations for life on a stony substratum; all the Plecoptera; and several Trichoptera. Among species encountered for the first time, there is another limpet, *Ancylus lacustris*, which lives on the stems of the reeds; several species of water-beetle and the water-boatmen *Corixa* and *Notonecta*, all forms which are dependent on the air for their oxygen; and *Enallagma* and *Coenagrion* (dragonflies), *Leptophlebia* (Ephemeroptera), and two stoneflies, *Nemoura cinerea* and *Nemurella picteti,* all essentially still-water forms.

A more recent study covered much of the west shore of the lake and of the east shore of its south basin. Fewer collections were made on the east side of the north basin, as the shore here has been modified by the owners of the gardens which bound it. It became apparent that the fauna of the stony sub-stratum is not uniform, as was thought likely at the outset. A distinct community at the north end of the lake is gradually replaced by another, which reaches its peak in the region where Moon worked, and then gives way to the first as the south end of the north basin is approached. The community found here occurs throughout the south basin. We have chosen to illustrate this change by selecting four stations (Fig. 17) rather than by showing the figures for all the stations, which has been done elsewhere. Station 1 is at Red Nab, a little to the south of Moon's area. Stations 2 and 3 lie on either side of Rawlinson's Nab, a promontory which juts far out into the south basin at right angles to the shore. Station 3 on the north side, that is the one sheltered from the prevailing southerly wind, harbours a fauna typical of the south basin. Station 4 lies near the outfall of the Bowness-Windermere sewage works.

At Station 1 there are two flat, stone-clinging Ephemer-optera, *Ecdyonurus dispar* and *Heptagenia lateralis*, and a swimming form, *Centroptilum luteolum*. Several species of stonefly occur but only two are included in Figure 17, these being the two that are typical of lakes and are not equally abundant in the inflowing streams. There is one caddis, *Poly-centropus flavomaculatus*, which spins a net and subsists on animals caught in it. All but the last are more abundant at Station 1 than at any other, and two are confined to it. Also generally at Station 1, sometimes in large numbers, are the next six species, all of which are widely distributed. Above them come a third group, comprising three flatworms, *Cran-gonyx* and *Asellus*, which are scarce or absent at Station 1.

Passing Station 2 for the moment, we note that at Station 3 the six insects at the bottom of the list are now almost all absent, and the five species at the top are abundant. Their

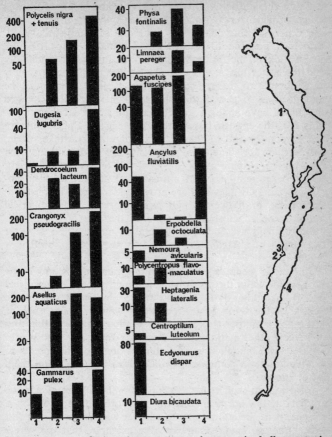

Fig. 17 Fauna at four stations on stony substratum in shallow water in Windermere

numbers are higher at Station 4 and therefore the conclusion that enrichment by sewage effluent influences their abundance is inescapable. It will be remembered that this community was found also at the north end of the north basin, a region under the influence of the only other large sewage disposal plant discharging into the lake. Both works are efficient and therefore the word enrichment, not pollution, has been used. Presumably the nutrients are utilized by some small organisms which provide a rich food supply for the flatworms, *Asellus* and others of this community, but this remains speculation. Nor is it possible to present an explanation supported by observation of the disappearance of the insects. It is unlikely that the water has any chemical or physical property that is unfavourable to them, and therefore a biotic effect must be sought; it is suggested that the animals which the changed conditions have enabled to become numerous eat them. Possibly it is the eggs that are eaten. The eggs of Ephemeroptera and Plecoptera fall to the bottom where chance takes them, and lie there for some time. Any consumer of debris, and even a herbivore, rasping away the algal felt on the stones, might take in large numbers. The animals that are successful in enriched waters lay eggs that are protected from casual consumption; those of flatworms and leeches are in large spherical tough cocoons, enormous relative to the size of the parent, those of snails are embedded in masses of jelly, and those of *Gammarus* and *Asellus* are carried by the mothers until they hatch.

Station 2, where species in both groups occur, might be supposed to lie in the north basin some distance in either direction from Station 1. In fact, as stated already, it was in the middle of the south basin in most of which a fauna such as was taken at Station 3 occurs. The peculiar feature of this, and of several other stations with a similar fauna, is exposure to wave action, though how this enables some of the insects to survive among a community which elsewhere eradicates them is not known.

Coniston, in which the effluent from the village affects the animals at the north end, is the only other lake in which a change from one community to another was observed. The rest were either so small, Esthwaite for example, that the whole lake was affected or so little enriched that there was no effect anywhere. If therefore the stations from the north end of Coniston are omitted and all the collections from each lake added together and expressed in terms of numbers caught in 100 minutes, comparison of the lakes may be made. The composition of the communities might be predicted from the findings in Windermere. In Esthwaite the number of flatworms collected is some four times as great as in the south basin of Windermere. Numbers diminish rapidly in other lakes, the nearer each is to the unproductive end of the series and none were found in Wastwater and Ennerdale. The same is true of the other species associated with the flatworms though the numbers do not fall in exactly the same way. *Gammarus*, for example, maintains a fair population in all lakes except Buttermere, Ennerdale, and Wastwater, where it is scarce or absent. *Physa* is absent from these three lakes, scarce in the next five, and fairly numerous in Coniston, Bassenthwaite and the south basin of Windermere, though numbers are lower in Esthwaite. The insects tend to be scarce, or even absent, in the productive lakes, to be most numerous in those near or beyond the middle of the series and then fewer again in the least productive lakes.

The most striking way to illustrate qualitatively the serial arrangement of the lakes is to record the number of species in three groups of animals that are not insects, which is done in Table 5 (p. 127). This table also includes Esrom in Denmark, a productive lowland lake. Preliminary surveys of the Cheshire and Shropshire meres indicate that they have a similiar fauna, characterized by abundance in both species and individuals of molluscs, and abundance of individuals of leeches, flatworms, *Gammarus* and *Asellus*. The flat Ephemeroptera nymphs, *Heptagenia* and *Ecdyonurus* are notable absentees on the stones in Esrom lake, and there is only one Plecopteron, *Nemoura avicularis*. The Corixidae also illustrate the point, though it is necessary to exclude those found in the middle of thick reed-beds, because they can be affected by local conditions. If a sandy bottom can be found in an unproductive lake, two species *Sigara dorsalis* and *S. scotti* occur. On a similar substratum in Coniston and Ullswater only *S. dorsalis* occurs, and in the most productive lakes it is joined by *S. falleni*. *S. dorsalis* is not known in Denmark, and its place is taken by the very similar *S. striata*.

Comparable changes in other groups have already been mentioned. The species of phytoplankton algae are different according to the position of the lake in the series, though the species of zooplankton are not. The proportion of the rooted plants, *Nitella* and *Isoetes*, relative to species of *Potamogeton* is high in unproductive lakes and decreases to a lower figure in productive lakes. The same is likely to be true of the animals inhabiting the bottom mud, for this is well known in other European lakes, and was indeed one of the bases on which they were orginally classified. At present, however, little is known about the inhabitants of this part of British lakes. The

lakes can also be arranged serially according to the species
of fish, though less well. Apart from the ubiquitous eel, trout
and char are the main species, in some possibly the only
species, of larger fish in the unproductive lakes. In the others,
perch and pike become increasingly important. In Windermere
small populations of roach and rudd maintain themselves, but,
as far as can be ascertained from the casual observations that
have been made, they are doing no more than that. In Esthwaite, in contrast, they appear to be increasing. Now, certain
defects in this otherwise regular picture must be mentioned.
Most notable, perhaps, is the absence of pike from Ullswater
and Haweswater. Another complicating species is *Coregonus*,
which occurs in Bassenthwaite and Derwent Water, where it is
known as the vendace, and in Ullswater and Haweswater,
where it is known as the skelly or schelly. It is another
salmonid which lives in deep water, though it comes inshore in
winter to spawn. They are also a species which probably spread
during the Ice Age and became isolated in various lakes when
the ice retreated. The four lakes named occupy the north-east
quadrant of the Lake District, which may be significant. However, any explanation of present-day distribution in terms of
the Ice Age events rests on the assumption that there have
been no changes in the intervening millenia, and particularly
that man has not moved fish from one water to another, an
activity to which he has been prone for the last thousand years
at least.

In Swedish lakes both *Coregonus* and char abound, and
workers there have shown that one sometimes replaces the
other. In Haweswater both occur together, and there is good
evidence that they did too in Ullswater until the last century,
when the char became extinct. The absence of char from Derwent Water and Bassenthwaite may be associated with the relatively shallow basins of those two lakes. It is also absent from
Esthwaite.

If all these data are put together in the way in which some
of it is presented in Table 5, there appear to be certain discontinuities which might justify the recognition of lake types. For
example, there are seven species of snail in Derwent Water and
the same number or more in lakes more productive, but not
more than four in lakes less productive (Table 5). However,
when such discontinuities are indicated for each group, they
are generally found to lie between different lakes, which is an

argument in favour of the serial concept and against classification into groups.

Little collecting has been done in the high natural tarns, but more is known about the artificial ones on the lower fells. Hodson's Tarn is an example (Fig. 8). It is a little less than half a hectare, which is just over one acre, in extent, and some three metres deep at the deepest point. It has one small inflow.

Table 5. Total numbers of species in three groups found in the Lake District lakes and in Esrom Lake, Denmark; and some indicator species

	Wastwater	Ennerdale	Buttermere	Crummock	Derwent Water	Bassenthwaite	Coniston	Ullswater	Windermere	Esthwaite	Esrom
Calcium mg./l.	2·4	2·2	2·1	2·1	4·5	5·3	6·1	5·7	6·2	8·3	42·0
Snails	2	2	3	4	7	7	7	7	10	8	19
Flatworms	1	1	3	2	2	4	5	3	5	5	7
Leeches	1	3	4	4	3	4	4	4	8	8	8

Typical *Sigara* of substratum with little organic matter:

	Wastwater	Ennerdale	Buttermere	Crummock	Derwent Water	Bassenthwaite	Coniston	Ullswater	Windermere	Esthwaite	Esrom
scotti	.	+	.	+	+	+	.	.	.	.	.
dorsalis	.	+	.	+	+	+	+	+	+	+	.*
falleni	.	.	.	.	.	.	.	.	+	+	+

* *S. striata*

It lies in an area where there are many tarns, the first of which was built to supply electricity to a house early in this century, when electric light was rare. Later the tarns were used for trout fishing and a hatchery to supply them was built at the foot of the fell.

It is known that Hodson's Tarn was dry in the late thirties, possibly for a long period, and this may have been why, when work started on it in 1955, it was found to be filled with luxuriant vegetation. It is believed that this is typical of an early stage of a tarn's development when the plants can take root in the underlying clay. At the inflow end in shallow water was a zone of *Carex rostrata*, growing on what had been bog before the valley was flooded. Elsewhere in shallow water there were patches of lakewort, *Littorella*. In the middle near the inflow, and probably rooted in the debris brought in by the one small stream was a large patch of the floating pondweed, *Potamogeton natans*. In most of the rest of the tarn there was a forest of milfoil, *Myriophyllum alternaeflorum*. This vegetation was mapped in 1955 and again in 1964. The choice of this year proved to be a stroke of fortune, for in 1965 the vegetation began to deteriorate. Between 1955 and 1964 both *Carex* and *Potamogeton* advanced towards the centre of the tarn but the striking change was that shown by the *Littorella*. All the patches had greatly increased their size, the plants were of unusual length, and they were growing unusually close together. In subsequent years *Littorella* retreated until it was almost confined to the zone in which it was exposed during times of drought. Owing to a leak in the dam the level fell about 30 cm. during long dry spells. In this zone the plants were neither as long nor as densely crowded as formerly. The milfoil disappeared almost completely leaving much of the tarn bottom bare, though areas were colonized by *Potamogeton alpinus*. This reduction in the vegetation was probably associated with the accumulation of poorly decomposed plant remains. Possibly the organic matter absorbed the nutrients that plants require and held them so firmly that none were available. There may have been a substance toxic to plants.

Nothing is known of the phytoplankton of Hodson's Tarn, but Dr J. W. G. Lund examined weekly samples from the somewhat larger Three Dubs Tarn. He produced a long list of species and showed that some of the abundant ones are numerous for a week or two only. How similar the sequence of events is in other years or in other tarns is not known. A thick growth of algae covers the rooted plants and, when the *Myriophyllum* was disappearing, plants seemed to be weighed down by the burden of algae upon them.

It is more instructive to take the animals according to their position in the food chain rather than according to their position in the animal kingdom. As is usual, little is known about the microscopic animals, though they probably play an important part in the economy of the tarn. Twelve species of Cladocera and 5 species of Copepoda are known from the tarn, and of these 3 and 1 respectively are probably truly planktonic. The rest live associated with plants, feeding on organisms attached to them.

The bottom mud is inhabited by chironomid larvae, oligo-

chaete worms, pea mussels (*Pisidium*) and larvae of the Alder
Fly, *Sialis*. The adult chironomids emerging from Three Dubs
Tarn were trapped one year, the traps being emptied once a
week. Over forty species were identified by Dr P. Freeman at
the Natural History Museum, and two were new to science.
Emergence periods were variable, some species coming out
steadily throughout the season, others during one or more
short periods. However, the time required to handle this catch
was so great that little was left to devote to other species, and
accordingly, when attention was turned to Hodson's Tarn, it
was decided to study the weed fauna only.

It is easy, if tedious, to discover what a herbivore eats, more
difficult to discover from what it derives its nourishment.
Many herbivores eat copiously and retain their food for a
short time only, with the result that some algae pass through
the gut unharmed. The commonest herbivore in Hodson's
Tarn, indeed by far the commonest of the macroscopic invertebrates, is *Leptophlebia* (Ephemeroptera). There are two
species, *L. marginata* and *L. vespertina*, with apparently the
same mode of life. However, *L. marginata* grows larger than
L. vespertina and emerges about a month before it. It is probably larger throughout life, and therefore the two species are
never dependent on exactly the same food at the same time.
The adults of *L. vespertina* are on the wing from mid May to
early June, the eggs lie unhatched for a month or two, and the
growing period is the winter.

Also belonging to the Ephemeroptera are *Cloeon dipterum*
and *C. simile*, which have different habitats, *C. simile* being the
commoner in the middle of the tarn. *Limnephilus marmoratus*,
a caddis larva in a cumbersome case of vegetable fragments,
and *Triaenodes* the swimming caddis in its neater case, the
Wandering Snail, *Lymnaea peregra*, and the Plecopteran, *Nemoura cinerea* are all herbivores. *Gammarus pulex* is said to be
more of an omnivore, and the two corixids typical of such a
tarn, *Sigara scotti* and *Hesperocorixa castanea* suck fine particles off the bottom which are found later in their guts in the
form of an amorphous mass containing little if anything that is
identifiable.

A rich variety of carnivores prey on these species and on
each other. There are those which can pursue their prey
actively, for example the trout, of which more later, and
dytiscid and haliplid beetles in both larval and adult stage.
Some of the larvae hunt for water fleas in the open water.

Another clearly distinct class are the lurkers, of which the dragon-flies are the chief example. They lie concealed in vegetation and seize passing prey, generally water fleas or chironomid larvae, with their hinged labium which, armed with two movable teeth at the tip, can be projected forward from beneath the head. Two Trichoptera, *Holocentropus dubius* and *Cyrnus flavidus* spin nets and probably feed in the same way as spiders on land. Less easy to classify are the caddis *Phryganea*, whose long case of vegetable pieces cut to size and glued together in a neat spiral, seems suited neither to concealment nor pursuit, and the leeches. Only five of the British leeches suck blood and the rest swallow their prey whole.

It is difficult to state exactly what the average carnivore eats during the year, an assertion that can be backed up by evidence from Hodson's Tarn, where the diet of the trout was studied. The food changes from day to day, probably from hour to hour, and according to the whim of the fish, and therefore samples of sufficient size taken sufficiently often would annihilate the population in too short a time. In Hodson's Tarn a sample of about ten fish was taken roughly once a week. In summer there had always been some feeding at the surface though no fish that had been feeding exclusively there was found. On wet and windy days when few animals might be expected on the wing, few aerial species were found in the trout. In a sample of ten it might be found that, whereas most contained two or three specimens of say *Lymnaea peregra*, one specimen would be packed with them. This specialization on one species occurred sufficiently frequently to make the calculation of an average a dubious procedure. Accordingly we regard with some scepticism the 'availability index' which some authors have based on the proportions of a species in the total population and in the food of trout. Such an index varies continually throughout the season. One generalization that does appear justified is that trout do not take very small items of food if large ones are available. It is therefore not unexpected to find that in Hodson's Tarn nymphs of *Leptophlebia* are not taken by trout till late in the year, two or three months after their appearance. Thereafter they form one of the main items of prey during the winter. *Lymnaea* is also eaten during the winter until the adults die after laying eggs in early summer. The snail then disappears from the trout's diet for a period, although numbers are at their greatest then. Nymphs of the small dragon-flies are eaten in larger numbers in the spring than at other times, perhaps because they are restless then as the time for emergence approaches. Trout evidently do not grub in the bottom of the tarn, and chironomid larvae are eaten only at the times when they come out of the mud to pupate.

The dragon-flies of the tarn merit more attention than they have received so far. There are seven species altogether, four large, and three small. The nymphs of the large ones are undoubtedly numerous in some years, to judge from the number of cast skins that are found beside the tarn after emergence, but in the tarn they lurk somewhere where the regular sampling brings in only a few. The three small ones inhabit the vegetation, all are numerous, and all abound in the samples. There are two contrasting types of life history. The eggs of *Pyrrhosoma nymphula*, the small red dragon-fly, hatch in late summer and the nymphs reach a length of 2 or 3 mm. before their growth ceases for the winter. In the following summer they reach full size; they pass a second winter in the water, and emerge in the following May or June. In 1957 the crop of new nymphs was exceptionally large. During the autumn and winter development was as usual, but at the end of the summer, when full size would normally have been attained, there were two groups. The nymphs in one were nearly full grown, those in the other had increased in length by no more than a millimetre or two. By the following spring the nymphs of the first group had attained full size and they in due course emerged as if they had belonged to a normal year. The nymphs in the group of small size, the larger group numerically, required a second summer to attain full size and emerged a year later than usual. It is suggested that in every generation a certain number find a feeding place and grow normally, while the rest, which are conveniently referred to as 'starvelings', are confined to places where they may survive for a long time but grow not at all. There is evidence that a few nymphs grow sufficiently fast in the year of hatching to emerge the following year, and Mr. A. E. Gardner reared this species from egg to adult in an aquarium in twelve months.

The life history of *Enallagma*, the small blue dragon-fly, is similar but less easy to make out because the rate of growth of each year class covers a bigger range and the identity of each class soon becomes less clear.

The life history of *Lestes sponsa*, the small green dragon-fly, is much less elastic than the ones just described. The egg develops to a certain point and goes no farther while subjected to low temperature. Then, when the temperature rises, development proceeds. The result of this obligatory diapause, to use the technical term, is to prevent eggs hatching until the spring. Growth is then very fast compared with that of *Pyrrhosoma* and *Enallagma*, and adults start to emerge in July. *Lestes* has a life history that is well adapted to the British climate, and it too is a successful species found abundantly in many places. Many species of dragon-fly are less successful to judge from their more limited distribution, but why they are less successful has yet to be discovered.

The nymphs of these three species are not distributed in the tarn in the same way. The most uniform distribution is that of *Lestes*, followed by *Enallagma*, which, however, is scarce in the *Carex*. *Pyrrhosoma* is abundant here and all round the tarn in shallow water but few nymphs are found in deep water. Straightforward observation explains this distribution. A pair of adult *Pyrrhosoma* remain together during egg-laying. They alight on some stem or leaf and the female immerses her abdomen as far as she can without wetting her wings. She feels around under water with the tip of her abdomen, inserts eggs into the stem or petiole and, this done, the pair fly off. Suitable landing places for this type of oviposition are found only in shallow water. The nymphs move little from the area where the parents alighted to lay eggs. *Lestes* also remain together, and both members of the pair crawl down a stem so that the eggs can be laid well below the surface of the water. They then crawl out of the water, the male still gripping the female firmly. The nymphs are much more active than those of the other two species and soon spread over the whole tarn. Pairs of *Enallagma* may be seen flying low over the surface of the tarn, keeping to those regions where there is nothing emerging high enough to impede this flight. The only landing places are the inflorescences of *Myriophyllum*. They project a millimetre or two above the surface and are evidently too small to attract *Pyrrhosoma* or *Lestes*. *Enallagma*, in contrast, does settle on one and the female immediately starts to walk into the water. The male flutters violently when his mate starts to submerge him, but eventually lets go and flies off alone. The female can be seen crawling down the plant, now and then doubling up her abdomen to pierce the stem and insert an egg. She may descend two or three metres and remain submerged for half an hour, but eventually she releases her hold and shoots to the surface like a cork, buoyed by the air bubble held between her wings. The distribution of the nymphs of *Pyrrhosoma* and *Enallagma* thus depends on the egg-laying behaviour of the parents.

One of the objects of studying a place such as Hodson's Tarn for a number of years is to find out how much the fauna varies from year to year, an important piece of knowledge for anyone wishing to compare different places. Total numbers of any one species have varied considerably but the list of species has remained astonishingly constant over the years. This is astonishing in view of the number of inhabitants that can fly at some stage of their life history. Even species that are always scarce turn up regularly.

Tarns of this kind are generally stocked with trout and it was thought that the close presence of this large predator might be an important ecological factor. Hodson's Tarn seemed to be a good place to investigate this, one of the few factors that can be studied in isolation from the others. Accordingly all the trout were removed and after five years the tarn was heavily restocked. When there were no fish, tadpoles could be seen swimming everywhere in the tarn with, one might say from an anthropocentric point of view, the utmost disregard for their personal safety. *Notonecta*, the water-boatman, which hangs from the surface waiting for prey to fall in and become entrapped in the surface film, might be seen all over the tarn. Beetle larvae swam freely through the water. All these disappeared when trout were introduced; they plainly lacked the behaviour that would enable them to live in the same piece of water as a large carnivore. The other species to disappear were the rarities. Species of corixid and beetle that had previously turned up year after year, were no longer recorded in the catches. *Hesperocorixa castanea* occurred everywhere in shallow water in the absence of fish, but, on their return, was confined to places where *Carex* and *Sphagnum* evidently provided enough cover to protect them against predation. Similarly *Lestes* disappeared from the *Myriophyllum* but survived in shallow water.

The abundant species were those that suffered least and such species as *Pyrrhosoma* and *Leptophlebia* were only slightly less numerous, at the end of a generation, when fish were present than when they were not, although large numbers were found inside fish. One possible explanation is that losses of large specimens, which were the ones which the fish took, were made good from the reserves of starvelings whose existence was postulated when the life history of *Pyrrhosoma* was described. If this explanation is correct, it must be supposed that, in the absence of fish, most of the starvelings eventually died. When fish were present a certain number were able to take over good feeding places whose former occupants had fallen a prey to fish and with it a rapid rate of growth. The result was that the human collector took nearly as many large nymphs when there were fish to eat them as when there were not, but in fact when predation was occurring the production of dragon-fly flesh was greater. The numbers of *Lymnaea* were higher when fish were present than they had been formerly, but after a few years the population fell suddenly to a low level. One possibility, now under investigation, is that one of the trematode parasites that infest snails died out when there were no trout because that fish is the alternative host. The newly introduced fish came from a hatchery and could have been free of parasites and remained so for a year or two. When the parasite found its way back to the tarn the snail population fell.

There are a few practical lessons to be drawn from this experiment. The animals important to trout as food inhabit weeds; mud fauna and casualties on the water surface provide supplementary rations in summer. Within the weeds the animals were safe, and it was probably only those that strayed too near the edge that were taken by fish. This would probably happen when the population was too large for the resources, which it probably generally was; in other words the fish were cropping a surplus and not affecting the total population. Any addition of animals to the tarn would have been profitless, but any step to increase the vegetation would have been useful. The way in which the rate of growth increased as the population of fish was reduced by the sampling bore out the scientific contention that most tarns of this kind contain too many fish.

Few comparable studies having been made, it is possible to do no more than offer a few notes on the distribution of species in some groups. In lowland productive ponds *Cloeon dipterum* is abundant and *Leptophlebia* absent. One may speculate whether *Cloeon*'s habit of laying eggs that hatch at once may not be significant in this context. The resulting nymphs can keep near the surface when the bottom becomes deoxygenated, whereas the eggs of *Leptophlebia* would be lying helplessly on the bottom during the time of year when this is most likely to happen.

Had there been a bigger area of exposed stones in the tarn, *Ancylus*, the freshwater limpet, might have been expected. Had the tarn been larger, or more calcareous, *Planorbis albus* would probably have occurred as well as *L. peregra*. Dr T. B. Reynoldson has brought together data that show that *Asellus* is rarely present in waters with less than 5 mg. calcium per litre and generally present in those with more than 12·5 mg. In concentrations in between it may or may not occur and the factor that determines whether it does or not is unknown. Davies (1969) presents evidence that flatworms would inhabit places such as Hodson's Tarn but for the heavy predation on them by dragon-fly nymphs. *Hesperocorixa castanea* and *Sigara scotti* are both indicators of extreme conditions. In more productive tarns they would be joined by *H. linnei* and *S. distincta*, both of which have been taken in Hodson's Tarn. In enriched tarns they might be replaced by *H. linnei* and *Callicorixa praeusta*. If the tarn were above about 500 m. (1,500 ft.) in the Lake District, *Callicorixa wollastoni* and *Arctocorisa carinata* might be expected to be the only species. Were it brackish, they would be *Sigara selecta* and *S. stagnalis*.

The obvious river on which to base an account of running water is the Duddon, surveyed, as already noted, by two American colleagues (Kuehne and Minshall 1969). It had long been known that Plecoptera tend to predominate in the head waters, but the extent to which it is true had never been based on figures. In the whole of the upper basin of the Duddon six species of stonefly nymph were abundant, and about eight more were recorded. *Pedicia rivosa*, a crane-fly larva, *Dicranota* of the same family, *Simulium* species, some net-spinning and free-living caddis, and the flatworm, *Crenobia (Planaria) alpina*, none of them numerous, were other animals found. Only in the lower basin and its tributaries were these stoneflies joined by the following familiar inhabitants of stony streams: *Ecdyonurus venosus, Heptagenia lateralis, Rhithrogena semicolorata, Baetis rhodani, Baetis muticus* (all Ephemeroptera) *Gammarus pulex, Ancylus fluviatilis*, various helminthid beetles and a number of Trichoptera.

If temperature were the important factor, the same fauna should be found in all the streams which are at a similar altitude. It is not. In Whelpside Ghyll, rising in Brownrigg Well, at an altitude of 850 m. (2,800 ft.) below the summit of Helvellyn, Plecoptera are numerous in the upper parts, but *Gammarus* occurs as well and is abundant in the spring. Several of the widespread stream species of Ephemeroptera occur, together with two others both characteristic of high altitudes, *Baetis tenax* and *Ameletus inopinatus*. It is less easy to be certain about other physical factors, but no evident differences between the various high streams have been recorded. Chemical differences may be more important. Analyses of a number of Lake District becks have been performed mainly in connection with the distribution of *Gammarus*, which is obviously a key species because, abundant in many streams, it is absent from others. Hynes did not find it in his classic study of the Afon Hirnant, and he records its absence from many other streams in Wales. *Gammarus* is unique among freshwater animals and, as there is only one species in the Lake District, it can be identified by those who, without any special knowledge, are anxious to find an excuse for exploring and camping in the high fells. Valuable contributions to a map of its distribution have, therefore, been made by such organizations as the Brathay Hall Exploration Group. *Gammarus* has been found in almost every stream flowing off the Helvellyn range and in many in the southern part of the region, but it is often absent elsewhere. Analyses suggest that there may be a correlation between calcium and *Gammarus* but there are several anomalous figures. If a definite correlation were to be established, there would still remain the question of how calcium acts. Do stonefly nymphs thrive in the upper basin of the Duddon because their requirement of calcium salts in solution is lower than that of *Gammarus* and Ephemeroptera, which do not? Or has calcium an effect on the food of such a nature that stoneflies can flourish on what there is to eat in the upper basin of the Duddon and the other animals cannot? At the present moment research on the uptake of ions is in fashion and some of it has explained distributions. That it does not always provide the answer has been shown by Reynoldson's work on flatworms, mentioned on p. 152. Hynes has suggested that the basis of the food chain in streams is the vegetable debris washed in, but he believes that stream animals cannot digest it. They derive their nourishment from the bacteria and fungi that are breaking the cellulose down. A comparative study of the digestive processes of Plecoptera and Ephemeroptera should prove interesting.

In the lower Duddon itself the fauna is a sparse version of

what occurs higher up and in the tributaries. The unstable bottom turned over by the water pouring down from the whole drainage area is probably unfavourable for all species. When conditions are less rigorous, the fauna increases as a stony river grows larger by the addition of species to genera that are represented higher up. *Rhithrogena semicolorata* is joined by *R. haarupi* (Ephem.), *Perlodes microcephala* by *Isogenus nubecul*, *Leuctra fusca* by *L. geniculata* (Plecopt.) and various helminthid beetles by *Limnius tuberculatus*. There are also some replacements: *Heptagenia lateralis* by *H. sulphurea*, *Ephemerella ignita* by *E. notata* and *Ecdyonurus venosus* by *E. dispar* (Ephem.) The last is the only one of all these species recorded in the Duddon.

The outflow of a lake carries plankton in abundance, a source of food not available in other parts of stony streams. As a result there occurs in such places an unusually dense population of net-spinning Trichoptera of the families Hydropsychidae and Polycentropodidae and also *Simulium*.

Ford Wood Beck is another Lake District stream which differs from most streams of the Duddon system in being more productive, because most of it runs through agricultural land. Ephemeroptera outnumber Plecoptera and *Gammarus* is abundant everywhere. A survey of it was made at a time when only four of the fourteen houses in a village through which it runs had indoor sanitation and baths, and in them the cisterns could be filled only by pumping. The waste water ran into a well-sited septic tank. When mains water was brought to the village, all the houses were connected to the same septic tank and there was some enrichment of the stream. The result was an enormous increase in the numbers of the flatworm *Polycelis felina* (*cornuta*). During the years of the original survey the total number taken in five minutes collecting at each of five stations was 2, 2, and 6. After the enrichment there were up to 1,000 specimens in identical collections. The large stonefly, *Perla bipunctata*, and *Ecdyonurus torrentis*, disappeared. The numbers of *Rhithrogena semicolorata* and *Baetis rhodani* fell. All these species live on the surface of stones rather than underneath them or in the gravel. It is known that flatworms can trap animals larger than themselves by laying threads of sticky mucus and overpowering anything caught in it by weight of numbers, and it was this method of hunting which probably caused the reductions observed. It cannot be the main one, for the flatworms are so numerous at some stations that they would annihilate the other macroinvertebrates if it were. No doubt, as in the lake, the enrichment caused the proliferation of some smaller organism which enabled the flatworms to increase in numbers, and then this large population reduced the number of certain insects by casual predation. The change brought about in the stream is much less than in the lake.

There have been recent studies of small stony streams all over Britain, but a detailed survey of a river, comparable to that of the Susaa in Denmark, remains to be made. Pentelow and Butcher and their colleagues laid the foundation for this work but it is now possible to name so many more species than they could that their lists are of limited value only.

When a river is no longer floored with stones but with sediment sufficiently fine for the roots of plants, many typical stone dwellers disappear. The flat nymphs of the Ecdyonuridae and *Ancylus* are no longer found nor are many of the stone-flies that live in gravel. In the weeds *Simulium* larvae are often very numerous. *Baetis vernus,* and other species too, join *B. rhodani* and *B. muticus.* Particularly in calcareous areas there is an abundance of both species and individuals of snails. As the river becomes slower the fauna becomes more and more like that of a lake or pond, though richer because the water brings a constant supply of nutrients.

The fauna of the upper Tees is much as might be expected from the findings of later workers to judge from the account given by Butcher, Longwell and Pentelow (1937). They record that the fishes are the trout, dace, chub, and minnow, which are common; the grayling, gudgeon, and roach, which are fairly common; and the pike and perch, which are rare. But only the trout and the minnow are found in the upper reaches, and the rest are confined to the lower, richer reaches. In the lower Tees the fauna is modified by pollution. An account of it is given in Chapter 14.

Some twenty-five miles from the mouth increased salinity due to sea-water first becomes detectable. Much of the fresh-water fauna stops abruptly, but a few species, the hog-louse *Asellus,* the worm *Tubifex tubifex,* the leech *Erpobdella octoculata,* and the snails *Lymnaea peregra, Ancylus fluviatilis,* and *Potamopyrgus jenkinsi,* penetrate for a greater or less distance. They meet a few marine animals which can live in dilute sea-water. The number of true brackish-water species, which occur neither in fresh water nor in the sea, is small (Fig. 18). There are two shrimps, *Gammarus zaddachi* and *Gam-*

marus duebeni, a flatworm *Procerodes (Gunda) ulvae*, and a colonial coelenterate *Cordylophora lacustris*. The absence of all four species of snail which occur in estuaries is unexpected.

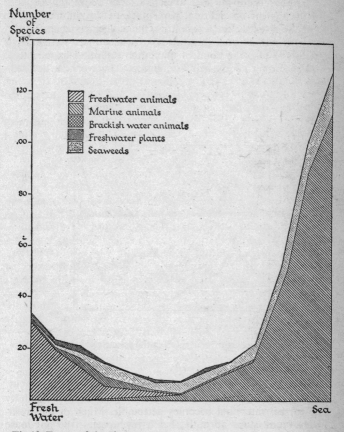

Fig.18 Fauna of the Tees estuary

There is also a brackish-water planktonic Copepod, *Eurytomora affinis* var. *hirundoides*.

Alexander, Southgate, and Bassindale (1935), who surveyed the estuary, noted that the number of different kinds of animals and plants diminished towards the middle of the estuary, and they summarized their findings in figures which are reproduced here as Figures 18 and 19. In Figure 19 the fresh- and brackish-water species are shown as a percentage of the number occurring in a typical stretch above the estuary, and the marine species as a percentage of the number in the sea at the river mouth; in addition, the oxygen concentration is shown. It will be seen that the point where there were fewest species was also the point of maximum pollution as indicated by the amount of oxygen in solution. The workers decided,

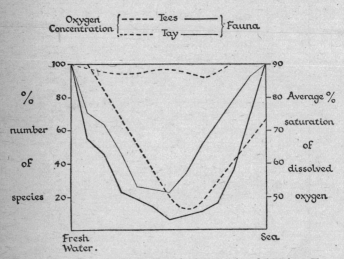

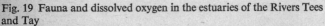

Fig. 19 Fauna and dissolved oxygen in the estuaries of the Rivers Tees and Tay

however, not to leap immediately to the obvious conclusion but to make first some observations on an unpolluted estuary, and they carried out a quick survey of the River Tay. The results are shown also on Figure 19, and it will be seen that there is in the Tay a diminution in the fauna towards the middle of the estuary just as in the Tees. Salinity changes are, therefore, the factor concerned and not pollution. But certain marine species penetrate into regions of lower salinity in the Tay than in the Tees and this may be due to pollution.

Studies of the fauna of water-courses are complicated by the movement of animals along it. The travels of *Crenobia* (*Plan-*

aria) *alpina* which moves upstream or down according to its physiological state are described in Chapter 11. *Gammarus pulex* tends to move upstream. Few *Rhithrogena semicolorata* were caught in an emergence trap near the head of a small stony stream, though nymphs were plentiful there, but many were caught in a trap near the mouth, which was interpreted as evidence of a downstream migration before emergence. Involuntary travel downstream may be caused by a flood, though the amount varies enormously according to conditions. Severe depopulation is recorded by some authors, no change in the numbers of animals by others. In addition there is a regular daily movement revealed by nets set so that the stream flows through them. When visited, they are found to contain most of the animals that inhabit the stream. Nearly all the species are few in the nets by day and much more numerous by night, often with a distinct peak soon after dusk.

It will be recalled that trout fry take up territories which they enlarge as they grow, with the result that there is a constant flow downstream of unsuccessful fish which have been squeezed out. They represent a surplus of production. To some extent the drift of invertebrates is comparable, for trout feed extensively on organisms they see in midwater, the numbers of a species drifting is greatest at the time when its growth is fastest, and the phenomenon does not depopulate the upper reaches. On the other hand, although many specimens are found being carried downstream, many of them do not travel far before regaining the bottom. There is also evidence of a steady upstream movement, particularly by small specimens. Some Trichoptera adults have been observed flying upstream to lay their eggs, but not all species winged in the adult stage do this.

Another estuary of a rather different type, for one of its main inhabitants illustrates a point which comes up for discussion later, is worth mention. It is the estuary of a small stream flowing over a pebbly beach in Wembury Bay near Plymouth. Pantin writes that just above the level reached by the highest tides there is a pool and this contains a fauna typical of fresh water, apart from one estuarine animal, the snail *Hydrobia ventrosa*. Between roughly the level of high-water spring tides and high-water neap tides there is a stretch where no animals were found, and Pantin refers to it as a 'desert'. Below the level of high-water neaps to a point about midway down the beach, where the shingle gives place to rock, there were only three animals, a flatworm *Procerodes (Gunda) ulvae*, a related creature which was not named, and a worm called *Protodrilus flavocapitatus*. In the rocky region they were joined by a few marine species and beyond the level of low-water neap tides they ceased to occur.

Brackish-water ponds have attracted attention. Howes (1939) surveyed in 1934 and 1935 a creek which had been dammed at both ends so that it no longer had connection either with the river of which it was once a distributary, or with the sea. The salinity was about two-thirds that of sea-water, but the relative proportions of some of the salts were not the same as in the sea. He found about fifty species. Most were marine animals known to be able to live in slightly dilute sea-water. There was only one freshwater animal, a beetle *Hydroporus planus,* but it was probably no more than a casual visitor, and there was no reason to believe that it bred in the creek. The rest was a characteristic brackish-water fauna; most of the animals had marine affinities but there were four related to freshwater species: *Phytia myosotis*, a snail, *Philhydrus bicolor*, a beetle, *Corixa selecta*, a water-boatman, and *Aedes detritus*, a mosquito. The last was rare, and scarcely to be expected in the open water of the creek, though it is abundant in brackish water in marshes and small pools.

Nicol (1935) investigated a series of pools in a salt-marsh. The highest were reached only by the highest tides and the average salinity was about 3% that of sea water, though after an invasion by the tide the salinity would rise to half that of sea water. Six species of water-beetle usually found in fresh water were fairly constant inhabitants of these pools, but the rest of the fauna was typical of brackish water: four Crustacea, including *Gammarus duebeni*, one mosquito *Aedes detritus*, and one beetle. During periods of neap tides this fauna might be augmented by freshwater species such as the mosquitoes *Culex pipiens* and *Anopheles claviger*, and the pond-snail *Limnaea truncatula*, but these would not survive an inundation by the sea.

The rest of the pools were entered by most tides. There was considerable variation in the fauna according to the nature of the substratum and the amount of vegetation. The commonest animals were brackish-water species with marine affinities, but a few creatures also found abundantly in the sea were common.

In inland waters of high salinity two salt-marsh mosquitoes,

Aedes detritus and *Ae. caspius*, may be abundant, together with various other dipterous larvae belonging to families about which much less is known. Other brackish-water animals which have been taken in inland localities are the shrimps, *Gammarus chevreuxi* and *G. tigrinus* and the two water-boat-men, *Sigara stagnalis* and *S. selecta*.

It may be remarked in conclusion that this is the chapter that has required most rewriting, but there is still an immense way to go before an account of all the plant and animal communities can be given. These words are being written in European Conservation Year (1970), and it is worth stressing that we do not know a great deal about what we are striving to conserve. Recently the Natural Environment Research Council was set up, and it is therefore to be hoped that the second half of the foregoing sentence will be among the first parts of this edition to require revision.

8. A Closer Look at the Environment

In this chapter we present such explanations as there are of why any species occurs where it does. This cannot be done until means of identifying the species can be found, a line of work which is not fashionable at the present time (1970). The larvae of many species of Trichoptera, for example, have still not been described, although caddis larvae abound in most pieces of water. The taxonomic hurdle negotiated, the ecologist can proceed to systematic observation in the field, either taking one group and studying the distribution over a wide area, or taking a whole community. This latter approach, on which the previous chapter was based, has not been popular because it involves the identification of species in so many different groups, but it is one which must be undertaken sooner or later because the other animals and plants are an important constituent of the environment in which any species lives. During the course of such work many plausible explanations of distribution will come to light. Most, however, will remain speculation until tested by some kind of experiment, either in the field, or, more generally, in the laboratory.

Certain resemblances between the communities of different environments have been noted; the stony region of a lake shore, for example, was said to be like the torrential parts of a stream, and a sheltered reed-bed in a lake was compared with a pond. These resemblances, however, are sometimes only general, and differences in detail become apparent on closer examination. *Ecdyonurus* (one of the flat mayflies), for instance, was noted as a form typical of and restricted to the stony lake shore and the torrential reaches of streams and rivers. But a closer examination, involving the identification of the species, shows that almost every specimen on the lake shore is *E. dispar*, and in the stream *E. torrentis*. *Notonecta*, the water-boatman, provides another example; specimens collected from a lake reed-bed are nearly always *N. glauca*; specimens from tarns almost invariably *N. obliqua* (*furcata*).

Every species has an optimum habitat, although the range of conditions in which it is found may be very wide. The wandering pond-snail, *Lymnaea peregra*, already mentioned, is a good example of a very adaptable species which can live and succeed in a great many different kinds of place. Most species are more circumscribed, and some are found only within a small range of conditions; the larvae of three British mosquitoes, for example, are hardly ever found anywhere except in holes which have rotted away in trees and filled up with rain-water. An experienced field naturalist can often tell at a glance whether or not a certain species is likely to occur in a certain place, but it is only rarely that he can analyse the habitat and point to some measurable feature with which the species is associated.

Many of the animals encountered in fresh water are the immature stages of airborne adults. If one of them is confined to a certain type of habitat, two explanations are possible; either the female chooses that sort of place to lay her eggs, or she lays her eggs at random and the resulting progeny survive only in that sort of place. Such meagre evidence as is available points to the former alternative, that is deliberate selection of the breeding place, as the more usual of the two. Other animals such as the water-beetles and the water-bugs can, in the adult stage, take wing and leave the water where they are. How far they select a new place to live is not known, but at least it can be said that they are not condemned to perish should they happen to alight in a place where conditions are not suitable.

But for the rest of the freshwater animals the possibility of selection of this sort does not exist. They may be carried passively on birds' feet, in naturalists' pond-nets, or in some other fortuitous vehicle, and if they arrive at a destination where conditions are suitable they may establish themselves and thrive, if not, they perish.

Suitable or favourable conditions have been mentioned several times. Presumably a habitat is suitable or favourable when a variety of physical, chemical, biological, and other factors fall within certain limits. But it is only rarely, in the present state of knowledge, that precise values for these limits can be given. The reason is partly because the range of one factor will vary according to the intensity of some other factor; temperature and rate of flow in the ecology of the flatworms *Crenobia alpina* and *Polycelis felina,* mentioned later on page 149, is a good example of this. When, as is probably usual, a number of factors interact in this way, it is a difficult feat to sort out the exact effects of each.

The ecologist is sometimes exercised to explain absence rather than presence. Absence may be due to events in the past, and anyone interested in distribution over a wide area, such as a continent, will devote much time to the Ice Age and what has happened since. Thienemann (1950) has written a standard work on the subject, and the distribution of every freshwater animal in a Europe divided into twenty-five regions is recorded in a series of recent publications edited by another professor from Plön (Illies, 1967). Britain became an island some 8,000 years ago, at which time the European fauna was advancing northwards from the regions where it had found refuge at the height of the Ice Age. A certain number of species had not reached Britain when the English Channel was formed, and only a few of these have succeeded in crossing the sea since. As a result we have a poorer fauna than our continental neighbours. Even within a small area such as the British Isles, the distribution of some species is related to past events.

During the Ice Age the land, in certain places, the Scandinavian peninsula being the classic example, was weighed down by the load of ice. When the ice melted at lower levels, the sea came in and covered areas which later became dry land as the ice cap gradually disappeared and the land, relieved of the weight, rose. Sea-water was trapped in depressions and with it a sample of the marine fauna and flora. These lakes lost their salinity in the course of time as freshwater streams, fed first by melting ice and later by rainfall, found their way to them. Most of the marine animals died but a few, possessed of greater powers of adaptability than others, survived and exist to this day. They have changed as time has gone by and are now distinct species, but they are obviously related to existing marine forms and have but slight affinity with the rest of the freshwater fauna. Two well-known examples of these marine relicts are a small shrimp, *Mysis relicta*, and a copepod, *Limnocalanus macrurus*.

Both occur in Ennerdale Water. As far as is known they do not occur in any other Lake District lake, though only Windermere has been investigated with a thoroughness sufficient to warrant a categorical statement. It is not easy to see how this distribution came about, and it could not have been through simple submergence in the way just described. There is no evidence that the ice of the Ice Age depressed the Lake District the necessary 368 feet (110 m.) to bring Ennerdale Water to sea level, and, in any case, had it done so, all the other lakes would have been submerged as well. One explanation is that, at a time when Ennerdale Water was dammed up by an ice barrier and was larger than it is today, a fresh advance of ice pushed up part of the Irish Sea and forced it into the lake. At that time valleys to the south were free of ice, and valleys to the north were frozen solid so that they did not retain this sea-water and its fauna, if it was ever forced into them.

However, the present problem is not how *Mysis* and *Limnocalanus* originally got into Ennerdale but why, in the intervening twelve thousand years or so, they have not spread to the other lakes. Certainly conditions in the other lakes are not identical, but it is improbable that two organisms which have managed to survive a change from sea-water to almost pure water are likely to be affected by the small differences, measurable in parts per million, which exist between the lakes today and which were probably smaller still in earlier times.

The most likely explanation is that these organisms have not an efficient means of dispersal. This is clearly a point which calls for close examination. Many freshwater habitats are quite isolated, and the rest are connected only by running water, which is not a suitable habitat for many pond and lake-shore forms. Therefore, before the absence of any species from a given habitat is attributed to some unfavourable factor in the environment of that habitat, it is necessary to examine the possibility that the species is absent simply because it has failed to get there. Almost all the insects have a winged adult stage and dispersal from one habitat to another presents no difficulties, but the other groups cannot live out of water for long. Professor A. E. Boycott kept under observation for many years the snail fauna of some 170 ponds near his home, and 69 of these were what he describes as closed, that is they had no permanent surface inflow or outflow. He found that, on the average, each pond received a new species every nine years. He could only detect a new arrival when it was a species not already present in the pond, and so the rate at which snails were entering each pond from outside must have been several times greater than this. How it came about was a different problem to which Boycott was not prepared to supply the answer.

Also in his area was a reservoir which harboured certain species not found in any pond. He transferred a couple of these to a number of ponds. One soon died out, and evidently there was something in the pond environment which was unfavourable to it. The other, *Planorbis corneus*, the great ramshorn, throve in nearly every pond into which it was introduced, and Boycott concluded that its absence from ponds was due to its inability to reach them. His general conclusion was that an inefficient means of dispersal was an important ecological factor for a few species of aquatic molluscs but not for the majority.

Comparable observations for other groups are lacking, and this is a field where further exploration is much needed. But the impression is that, in general, Boycott's conclusions with snails are widely applicable. We conclude that the absence of any species from an old well-established and apparently favourable locality is only rarely due to the fact that the species has not succeeded in reaching it.

How freshwater organisms get from one piece of water to another is still mysterious. That many travel on birds' feet is indubitable, but it hardly seems likely that this means of transport is sufficient to account for all the traffic. Some have eggs or spores that can withstand drying-up.

Fig.20 Distribution maps of three British Corixids (water-bugs): *a. Micronecta scholtzi*; *b. Callicorixa wollastoni*; *c. Sigara scotti*

In general then species are absent from parts of a small area such as the British Isles because conditions are not suitable for them. Some clue to what makes them unsuitable may be gained from distribution maps, such as those of three species of water-bugs shown on Figure 20. They are based on counties, subdivided when large, not on 10 km. squares as was the outstandingly successful *Atlas of the British Flora* (ed. Perring and Walters, 1962). Their success, however, was due to the large number of competent botanists available: 3,000 volunteered, 1,500 sent in cards, and 250 did most of the work. For a group such as the water-bugs there would be no more than a handful of volunteers and the resulting map would show more about the distribution of collectors than collected.

the distribution of collectors than collected.

Micronecta scholtzi (Fig. 20a) occurs in all the countries bounding the western end of the Mediterranean and northwards to Belgium and Germany, a distribution which suggests that it is confined to the south and east of England because the rest is too cold for it. The region in which it occurs is also one of low relief and generally calcareous soil, but the distribution in other countries indicates that temperature is likely to be the important factor. *Callicorixa wollastoni* (Fig. 20b), in contrast, occurs in north and west England and is in fact found only at high altitudes. Abroad it is known from Scandinavia and north Russia (Illies, 1967). Whether it is restricted to high altitudes and latitudes because elsewhere it cannot tolerate the temperature, or because it cannot compete with other species, is not known. *Sigara scotti* replaces *C. wollastoni* at about 500 m. (1,500 ft.) in the Lake District and the distribution pattern is not dissimilar. It is, however, widely distributed in the south of England particularly in those counties where there is much heath. It is associated with waters in which calcium is scarce.

It is now time to discuss some group where the limitation of range by temperature is a proven fact and not merely a probable hypothesis. The stream-dwelling planarians or flatworms provide a clear illustration. Two Balkan species, *Planaria montenegrina* and *P. gonocephala*, occur together in most streams but Beauchamp and Ullyott (1932), in the course of an extensive search, came across a few streams where only one species was present. When *P. montenegrina* was alone, it extended from the spring down to a point where the temperature was 16° to 17° C. When *P. gonocephala* occurred alone its range also went where the temperature was 21° to 23° C. When the two occurred upstream to the spring, but downstream it reached a point together, *P. montenegrina* occupied the upper and *P. gonocephala* the lower reaches, and there was little overlap where they met. The temperature at this point was 13° to 14° C, so that when the two species came into competition the range of each was curtailed.

In Britain *Crenobia (Planaria) alpina* and *Polycelis felina* occur together in a similar way, *Crenobia* from the spring down to a point where the temperature is about 13° C and *Polycelis* from here down to a point 3–4° C warmer. The temperature of a spring may be constant throughout the year but the range of diurnal and annual fluctuation increases with distance from the spring. It is therefore difficult to define precisely in terms of temperature the point at which one species gives place to another. On the continent a third species, *P. gonocephala*, whose lower limit has already been mentioned, occurs below *P. felina*. Competition between these species is likely, but it has not been demonstrated as clearly as that between the Balkan species. Sometimes *C. alpina* and *P. felina* occur together up to the spring and this is generally seen where flow is not rapid.

This distribution has been known for many years and much controversy has centered round it. Reliable information has been accruing over the years, but recent researches by Dr E. Pattee in France have done most to provide an explanation. First he established the lethal temperature on a criterion more exacting than that set by anyone else. He kept the flatworms in the laboratory and defined the lethal level as the temperature at which births exceeded deaths. The temperature was raised 2·5° C daily and held steady when the level to be tested was reached. At 25° C (77° F) the average time of survival of *C. alpina* was two and a half days. It increased to about sixteen days at 20°C and sixty days at 17·5° C. At 15° C two specimens out of ten survived to breed. Pattee concludes from this that the lethal temperature for this species lies between 12 and 14° C. Testing the others in the same way, he obtained the following results: *Crenobia alpina* 12–14° C, *Polycelis felina* 16–17° C, *Planaria gonocephala* 20° C, *Polycelis nigra* 23–24° C. The last is generally found in stagnant water, though Berg took it in the Susaa (Chapter 3).

Pattee next investigated the rate of growth and reproduction at various temperatures. *C. alpina* breeds below 5° C and the rate reaches a peak, a low one compared with the others, at 10° C. *P. felina* starts to breed at about 5° C, but the rate increases more rapidly than that of *C. alpina* and overtakes it at about 7° C, going on to reach a maximum at 15° C. The growth and reproduction of *P. gonocephala* is similar, but any given rate is found in water 2–3° C warmer, the peak, almost identical with that of *P. felina*, occurring at about 18° C.

Pattee concludes that the replacement of one species by another is due to the more rapid growth and reproduction at increasing temperatures down the length of a stream and that spread upstream is prevented by competition. If this were the whole explanation, *C. alpina* should be confined to a shorter and colder zone than it generally is, because *P. felina* is reproducing at a faster rate above about 8° C. An explanation of this anomaly was provided by Pattee's final investigation, which showed that *P. felina* avoids, and is swept away if it fails to avoid, a current which *C. alpina* withstands without difficulty.

In Ford Wood Beck the numbers of *P. felina* are generally lowest by a wide margin at the station in the swiftest reach.

Both *C. alpina* and *P. felina* are confined to streams in Britain because only water flowing from springs is cold enough for them. High in the Alps they are found on the stone shores of lakes.

The literature on flatworms is immense, but some papers are more important than others. One such on flatworms denied that *C. alpina* is a cold-water species on the grounds that it was active in an alpine pool in which the temperature was well into the twenties. This high temperature was reached only in the afternoon and the pool was cold at night. Some German workers, whose figures incidentally agree closely with Pattee's, thereupon undertook an investigation of tolerance and found that *C. alpina* could survive 25° C provided it was not exposed to it for too long. This work, however, attracted less attention.

Professor A. G. Dahm has shown that populations of *C. alpina* vary from stream to stream, chiefly in the temperature at which they breed most prolifically and the kind of reproduction, sexual or asexual, at different temperatures. Presumably *C. alpina* was widespread immediately after the Ice Age, but, as waters warmed up later, was split into isolated populations at the top of each tributary. It is possible that as time goes on each one will be recognizable as a distinct species.

The trout is another cold-water species, which dies if the temperature remains long above about 25° C. (Frost and Brown, 1967). There are probably not many waters in Britain which lack trout because they become too warm, but farther south in Europe the range of the species is limited in this way. The trout spawns in winter when the water is cold, and low temperature appears to play some part in stimulating reproduction. It is possible that some places are not too warm to prevent survival, but never cold enough for reproduction, and that this is the way in which temperature limits range. Conversely a species whose optimum temperature is relatively high may be kept out of certain waters because the water is not warm enough for reproduction sufficiently often. For example the late Dr D. S. Rawson observed that in two years out of four the breeding of an American fish introduced into Canada was abortive because the temperature fell below a level at which the males deserted the nests and left the eggs to die. If the rate of reproduction of an animal is slow because the water is cold, it may not be sufficient to make good losses caused by predators. Alternatively the species may not be able to compete with a rival whose optimum temperature is different; the flatworms mentioned above illustrate this. It must also be stressed that, as water becomes warmer, it holds less oxygen, and the requirements of organisms for that substance increase. It is therefore unwise to try to separate the two when discussing the ecology of animals. In fact ideally these various factors should not be taken one by one; rather the occurrence of species should be discussed in terms of all the factors that bring it about, for nearly always it is the result of the interaction of several. This, however, is a counsel of perfection in the light of present knowledge, because the reasons for the distribution of any one species have rarely been discovered. The outstanding exception to this generalization are the lake-dwelling flatworms, and it is logical, therefore, to take them next.

There are four common ones: *Dendrocoelum lacteum,* *Dugesia* (*Planaria*) *lugubris, Polycelis tenuis,* and *P. nigra.* Dr T. B. Reynoldson of Bangor, who, at the head of a school, has studied these animals for many years, observed that all four occur abundantly in calcareous lakes, and that the number falls steadily when lakes are grouped in classes with diminishing quantities of calcium. (The figures quoted in the previous chapter for Lake District lakes (Table 5, p. 127) bear out Reynoldson's findings well.) The obvious conclusion, particularly to a generation of biologists brought up to think in terms of physics and chemistry rather than natural history, is that soft water is a medium directly unfavourable to some of the species. Reynoldson showed that it is not. He had access to a llyn with remarkably soft water, some 0·5 mg./Ca. per litre, and he showed that all the species could thrive in it. Few adults kept a long time in cages in the pond itself died, and young could be reared easily in water from it in the laboratory. The Bangor school devoted a lot of time to the food of flatworms, a difficult study since flatworms can digest prey outside the body. The diet must therefore be investigated by means of a tedious but effective precipitin technique. Rabbits are sensitized, each one to some animal that a flatworm might meet in the wild. Then a flatworm is crushed and treated with a drop of blood from each rabbit. A precipitation indicates that the flatworm had eaten an organism of the species to which the rabbit yielding the blood had been sensitized.

In this way it was found that there are consistent differences in the prey of each species. *Dendrocoelum* eats more crustaceans, particularly *Asellus, Dugesia* more molluscs than the others. Now *Asellus* is abundant in productive calcareous lakes, and absent in those that are less productive and have softer water. Molluscs are fewer in both species and individuals down a series from one to the other. Therefore, towards the lower end of the series, what Reynoldson calls the 'food refuge' of *Dendrocoelum* and *Dugesia* is steadily diminishing, and all four species come into ever fiercer competition because all are relying on the same food. In this struggle *Dendrocoelum* and *Dugesia* are not successful. *P. nigra,* the survivor and *P. tenuis* differ in the kind of worm on which each feeds most readily. Davies (1969), also of the Bangor school, has made some experiments which show that *Pyrrhosoma* and newts prey on flatworms probably to an extent sufficient to eliminate them from a place in which they, the predators, are numerous. On the stony substrata of lakes, where flatworms abound, these two predators do not occur.

It is thus possible to state with some confidence why flatworms occur where they do both in streams and in standing water. No other group is comparable, and it is necessary to revert to taking the various factors one by one and describing those studies which have revealed how each affects the distribution of some animal.

Oxygen is an important factor but a difficult one to study because its concentration can vary so much in a short period. In most waters there are some plants to produce oxygen in sunlight; after dark they respire and consume oxygen, the concentration of which may be reduced considerably if the similar process of decomposition of plant and animal remains is proceeding rapidly. If an animal does not occur in a place because there are occasions when there is not enough oxygen for it, those occasions occur at about dawn, which is a time when biologists are not often out and about taking samples.

Knowledge has been advanced considerably by a Canadian school led by Professor F. E. J. Fry, who devised an apparatus for measuring the oxygen consumption of fish both when they were quiescent and when they were travelling at top speed. All too frequently a scientist invents a neat piece of apparatus for a certain purpose, obtains his results and then passes on to another problem. One of the reasons why the Canadian results are so noteworthy is that, once the apparatus was established, a number of fish were investigated. The resulting comparisons are of great value.

An animal working hard is conveying oxygen from the

medium to the tissues as fast as it can, and generally the amount reaching the tissues is limited by the transport system within the animal, in other words the gills cannot take up a greater quantity, the heart cannot beat faster, or interchange at either end cannot be speeded up. The oxygen concentration in the medium can fall from saturation without affecting the performance of the fish. Then a level is reached when that is no longer true; bodily activity is limited by the amount of oxygen in the water, not by the limitations of its system for transporting oxygen. Any further decrease in oxygen concentration will lead to a decrease in activity until a point is reached where the fish can stay alive only by reducing essential movement, such as passing water over the gills, to a minimum. The level at which the change takes place is called the incipient limiting level, and it varies greatly according to temperature and according to the species of fish. The trout is one of the most exigent of the species examined. At 15° C. the incipient limiting level is reached while the concentration of oxygen is still more than 75% of the saturation value, and at higher temperatures values a little below saturation begin to incommode the fish. Even at a temperature of 5° C. this takes place when the water is still more than half saturated with oxygen. A great contrast is provided by the goldfish which, even at temperatures as high as 35° C., is still moving as fast as its bodily limitations permit when the oxygen concentration is below 25% of saturation.

The ecologist cannot produce information to match in precision these findings of the physiologist. How often does a trout fail to elude a predator or to capture a piece of prey because the oxygen concentration is preventing it from attaining its maximum speed? There is no answer to these questions today, there probably never will be, and indeed it is impossible to say that oxygen is the important limiting factor. The trout is a species adapted to live in cool water with plenty of oxygen. Such conditions tend to be associated with running water or wave-beaten shores and a substratum of stones or gravel. It is only on the latter that a trout will lay eggs. It is a highly successful species in this environment and it is pertinent to wonder what powers of adaptation it possesses. If suddenly some catastrophe left it the only fish in Britain, would it, over the years, adapt itself to colonize water from which it is at present barred by high temperature or low concentration of oxygen, and would it change its breeding habits so that it could reproduce in places where there were only weeds or mud? It may be long before we find out, because even under experimental conditions, such a change might take longer than the span of human life. Many species have a big potential for adaptation, the outstanding example of recent times being the flies and mosquitoes which can now be killed only by a concentration of DDT many times stronger than that which sufficed to kill their ancestors. One could quote also those birds that have adapted themselves to life in towns. It might, therefore, be temperature, oxygen, requirements for breeding or the competition with other species that brings about the present distribution of trout. Probably the importance of each one varies with time and place.

This discussion sprang from an examination of the oxygen required for greatest effort. There are equally exact figures for the concentration at which life is just possible for a resting animal (level of no excess activity in physiological parlance), and the contrast between trout and goldfish is just as great. At 5° C. the goldfish requires only 0·3 milligrams of oxygen per litre which means that it could survive in heavily polluted conditions. The trout requires 2·3 mg./l. at the same temperature. Near its lethal temperature the trout requires 4 mg./l. which is nearly half the concentration of oxygen at saturation level and the goldfish needs only 1 mg./l. Here again the exact figures of the physiologist cannot be applied with the same precision by the ecologist. Obviously if the concentration sinks below the critical level, fish die, but by how much it must be exceeded in order that a fish may be sufficiently active to secure enough to eat, to grow and to reproduce, and to do these things in the presence of a species more tolerant of high temperature and low oxygen concentration, is not known.

Another important contribution to knowledge about oxygen as an ecological factor has come from Switzerland, where Dr H. Ambühl has studied the importance of flow. It is not important to a fish, because the fish creates its own current through the mouth and over the gills. A similar mechanism is possessed by, for example, two of the flat mayflies, *Ecdyonurus* and *Heptagenia*, as Ambühl showed. They possess what are commonly called gills, thin flat plates seven in number on the sides of each abdominal segment. They are like a series of short broad paddles but by moving them the animal does not progress, but causes water to flow over the body. This appears to be the main region where oxygen is absorbed, and the result of this ability to create a current is that oxygen consumption is at the same level whether the nymph is in still or running water. A relative, *Rhithrogena*, has adapted its so-called gills to form a sucker with which it can cling to a flat surface, and therefore cannot use them to create a current. In an experiment in which current speed could be progressively reduced, the consumption of oxygen was found to fall. This is likely to be why *Rhithrogena* is confined to running water, whereas *Ecdyonurus* and *Heptagenia* are abundant in running water and on the stony substratum of lakes. Similar to *Rhithrogena* is *Baetis* which has lost the power of making effective movements with its gills. In Britain the genus is confined to running water, but in north Scandinavia, where no doubt the low temperature reduces the demand for oxygen, a species occurs in lakes and smaller pieces of water.

Another important chemical factor in fresh water is the concentration of calcium. Botanists have long recognized that of the land-plants some can be classed as calciphile or lime-loving and others as calcifuge or lime-shunning. Well-known examples of the latter are bracken and bogmoss. Bracken, for example, which grows with such unwanted luxuriance on poor pastureland, is only with difficulty cultivated in the calcareous soil of the Cambridge Botanic Garden. Probably many animals can be assigned to the same two categories. Boycott recognizes a number of calciphile species of snail, and finds that the aquatic ones hardly ever occur in waters with less than twenty parts per million of calcium. A survey of the water-bug fauna of dew-ponds on the chalk-downs revealed that the three species common there were all rarities in the calcium-poor waters of the Lake District. Boycott is doubtful whether any species of mollusc can truly be called calcifuge though the pearl-mussel, *Margaritana margaritifera*, is a possible example. A number of species are found in lime-poor waters but they are not confined to such places, and Boycott believes that their scarcity in localities richer in lime is due to the fact that they cannot compete with the calciphile species. *Corixa scotti*, which has already been discussed, is comparable.

Little is known about how calcium acts. It certainly plays an essential part in some physiological processes. A well known example is the flatworm, *Procerodes (Gunda) ulvae*, which inhabits the inter-tidal parts of stony streams, where it is in fresh water at low tide and sea water at high tide. It can survive in these difficult circumstances because it can tolerate changes in the composition of its body fluid. It is able to damp down the fluctuations so that they are never as great as in the surrounding water, but an excessive uptake of fresh water by the more concentrated body fluids can be prevented only when the concentration of calcium is relatively high. In soft water the animals swell and burst. It seems likely that other animals, particularly those that have invaded fresh water from the sea, are ultimately limited in the same way; in the very soft waters of mountain lakes and tarns they cannot maintain their body fluids at the concentration essential for proper functioning. However, Reynoldson's work, described above, showed that this is not true of the common lake flatworms; they survived in a water whose calcium content was so low that waters with less must be rare. The water that comes from the sky as rain is not pure. Storms whip up spray from the sea and carry it far inland; chimneys contribute to the atmosphere a variety of substances which dissolve and descend in rain. His findings, and those of other workers too, are that the concentration of calcium gives a general measure of productivity. This is probably connected with W. H. Pearsall's observation that the decomposition of vegetable organic matter proceeds more rapidly in the presence of calcium, but it is a field of which exploration has reached no more than the fringe.

Many workers have analysed waters in the hope of finding differences with which faunistic differences might be correlated, and almost all have been disappointed. Savage (1961) found that frogs tend to lay eggs in ponds with more than the average amount of potassium, but generally this line of research has been profitless. It seems unlikely that organisms are uninfluenced by what is dissolved in water, and the failure to show that they are calls for examination. Sodium, calcium, magnesium, potassium, hydrogen, bicarbonate, chloride, sulphate, and nitrate contribute 99% of the ions dissolved in Lake District waters, and probably most others as well. These are the substances whose concentration is generally measured in any investigation of a piece of water. However, Lund has shown that, apart from nitrate, these substances change little in concentration as algal populations wax and wane. Though to the chemist they appear to be present in minute amount, generally around one or two parts in a million, the concentration is enormous relative to the requirements of the algae. Lund has shown that silicate limits the numbers of *Asterionella* in Lake District lakes, and other substances such as cobalt, boron, and manganese, generally present in such small amounts that they are known as trace elements, have been shown to limit the number of other algae in other places (Goldman, 1966). It may be that these trace elements affect some animals directly, or influence the composition of animal communities indirectly through their influence on plants. Unfortunately their detection in water requires resources only rarely available.

The numerous analyses that have been made have confined themselves not only to the common solutes, but to inorganic chemistry. When organic substances can be isolated and identified, some may prove important in the study of animal ecology. An investigation carried out in Assam by Dr R. C. Muirhead Thomson may be described because it illustrates not only this point but also others, notably that in ecology the obvious explanation is not always the right one. When Muirhead Thomson arrived in Assam in the late thirties, it had been known for many years that the vector of malaria there is *Anopheles minimus*, whose larvae occur in streams, and that species that occur in nearby borrow-pits are harmless. It was also known that *A. minimus* disappeared from streams over which a bush with particularly dense growth was planted, and further that the distribution of larvae is brought about by selection by the ovipositing females. Ross's discovery in 1898 that malaria is carried from man to man by mosquitoes was followed rapidly by the discovery that only *Anopheles* is the vector and then that only a few species in that large genus transmit the disease. This was a stimulus to work on the ecology of *Anopheles*, and it was soon known that each species is confined to a fairly well-defined type of water. Eggs or larvae transferred to the sort of place in which they did not usually occur generally throve.

This was the background from which Muirhead Thomson started. He looked at temperature, for it seemed likely that the stagnant borrow-pits in full sunlight reached a higher temperature than the streams with their partly shaded banks. Not only did measurements show this to be true, but experiments revealed that larvae of *Anopheles minimus* died at a temperature reached in the ponds. This high temperature, however, was reached by day, when the sun was up and the mosquitoes resting in some damp cool place. At night when they came out to lay eggs, both stream and pits were at much the same temperature. Clearly this was not the factor that took the female to streams to lay eggs and, as these could be identified, it was soon shown that random oviposition and elimination of eggs laid in pits by the high daytime temperature was not the explanation of the larval distribution.

Observation of the mosquito laying eggs produced some more surprises. It was absent from streams under dense shade, yet selected shaded places to lay its eggs. The larvae occurred in streams yet the female placed her eggs on water that was quite still. The absence of the latter was the explanation of the paradox. Under the bushes all vegetation died, and there were no longer leaves and stems trailing in the water and damming off the little areas of still water, which was what the female sought as a site for her eggs.

The physical factors of temperature, shade, and flow having been eliminated, Muirhead-Thomson turned his attention to chemical differences. He found that in general there was three to four times as much organic matter in the pit water as in the streams, with some slight evidence that there was a difference in quality as well. This was the point at which several other investigations had stopped. A female ready to lay eggs responds to several stimuli but it seems likely that as she gets close to the water her choice depends on whether or not it emits a particular odour, presumably that of some organic substance. But nobody has isolated such a substance and demonstrated its attractiveness to a particular species. There are, as already mentioned, three species of mosquito in England of which the larvae are found only in rain-filled rot-holes in trees. One is rare and the other two are to be found in almost every tree-hole. The number of records of any of the three breeding in any other sort of place are extremely few, and the number of records of other species breeding in tree-holes are even fewer. Many tree-holes are in a cavity with a small hole as the only means of access, and offer rather peculiar physical conditions; but this is not true of all, for some are in the stump of a tree which has been cut down and are no more enclosed and shaded than any other small body of water. It is unlikely, therefore, that physical conditions are of paramount importance in attracting the tree-hole-breeding species. The most plausible explanation, in the light of present knowledge, is that the mosquitoes are attracted by some specific organic substance which is present in water contained in a tree cavity and not in other freshwater localities.

Of the two subjects to which mosquitoes have led, the selection exercised by an ovipositing female may be deferred, and organic matter considered immediately. Organic matter accumulates on the bottom of a lake as plants die, as already described, and there can be two successions, one leading to fen, the other to bog. The species of corixid found depends on the amount of organic matter in the soil as is shown in Figure 21.

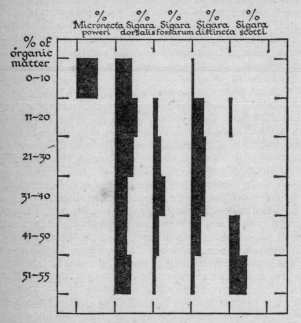

Fig 21. Succession of Corixid species with increasing percentage of organic matter in the soil of Windermere

On gravel or sand with little organic matter *Micronecta poweri* occurs. *Sigara dorsalis* also occurs where there is little organic matter, but requires some shelter such as is provided by a reed-bed. The lakeward edge and most of the middle of a reed-bed generally have a sandy bottom, and it is only towards the land that organic matter accumulates. Therefore the habitat of *S. dorsalis* is much larger in area than that of any other species and it attains an abundance which enables it to invade all the other habitats. In a tarn (Fig. 22) it is *S. scotti* which is numerous everywhere for the same reason. Two other species, *S. fossarum* and *S. distincta*, constitute a higher proportion of all corixids as the amount of organic matter increases to between 20% and 40%, and then a diminishing one as *S. scotti* comes in. The same species are found in the tarn except for *Micronecta* at the beginning, and *Hesperocorixa castanea* at the end. This is the succession under conditions that lead to bog. In Esrom Lake and in some Shropshire Meres *S. dorsalis* (in Esrom *S. striata* which is very similar) and *S. falleni* occur where vegetation is not too thick, *H. linnei* occurs in thick reeds and the

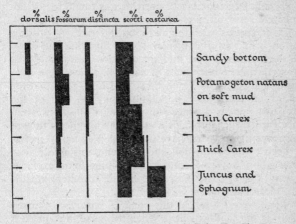

Fig. 22 Succession of Corixid species in Wise Een Tarn

succession culminates in *H. sahlbergi. S. fossarum* is associated with this succession. *H. linnei* is not abundant in any Lake District lake, but where fen has formed at the mouth of a stream *H. sahlbergi* is the species found in pools. In Esthwaite, the most productive of the lakes, *S. dorsalis* and *S. falleni* are found in equal abundance.

It is clear that the occurrence of each species is related not only to the total amount of organic matter but also to its nature, but little is known about how this comes about or of the individual substances of which the organic matter in various places is composed.

Dr E. J. Popham has shown that in several species the intensity of pigmentation may vary within rather wide limits and is influenced by the background on which the young develop.

In nature specimens are only rarely found on a bottom with which they do not harmonize in colour intensity, and, if they are placed on such a bottom in captivity, they appear ill at ease and may attempt to fly away. Furthermore, if fish be present, they devour corixids which do not match the background in greater numbers than those which do. These observations, however, do not provide a complete explanation of corixid distribution, because there may be two bottoms of the same colour intensity but different nature, and they will be inhabited by different species of water-bugs.

An unusually large number of corixids was found in a steep-sided concrete reservoir, and, in several successive collections, the abundant species was found to be different. A possible explanation is that swarms of corixids on the wing had dropped into the water and then, finding conditions unsuitable, had moved on. Various other pieces of water in which there were many species appeared to be unstable, which suggests that the presence of an established population may be one of the features which makes a place 'unsuitable'. Collections from Hodson's Tarn, which provides a stable environment, though extending over many years, have not yielded as many species as have been taken in one collection in some unstable places. This observation, however, contributes nothing to an explanation of why *S. scotti* and *H. castanea* and no others are the abundant species in Hodson's Tarn.

When there were no fish in this pond *H. castanea* occurred everywhere in the shallow water. When fish were introduced it disappeared from regions where there was *Littorella* and was found only in thick *Carex* and *Sphagnum*, a distribution similar to that in other tarns. It was found inside fish, when they were first introduced, and presumably they eradicated it from all areas except those where the cover was too thick for them to hunt. Several other species that had occurred regularly every year, though in small numbers, were not taken after fish were introduced, at least not by the human collectors. Several were found inside fish. Popham's work suggests that fish may have got them first because they were restless in surroundings in which, to put it unscientifically, they did not feel at home, but the question of why they did not feel at home remains unanswered.

It would, to quote the words of Professor Boycott, 'be unconventional beyond the bounds of propriety to make no mention of hydrogen-ion concentration'. It is commonly denoted by the symbol pH. At pH 7 a substance is neutral. Below pH 7 hydrogen ions are in excess of hydroxyl ions and conditions are acid, and above pH 7 the reverse obtains and conditions are alkaline. pH at one time enjoyed a great vogue among ecologists, partly because it is very easy to measure and a reading can be obtained in the field by mixing a drop of 'indicator' with a little of the water to be tested and matching the resulting colour with a set of standard tubes. Its popularity was enhanced by a certain worker who discovered a tree-hole with a pH of 4·5, a degree of acidity which the general run of fresh-water habitats rarely exceed. As a result of some not very critical experiments he reached the conclusion that the tree-hole-inhabiting species of mosquito were unable to survive except in an acid medium. Shortly after the publication of these results, two widely read text-books were written, and both referred to them as an example of pH as an ecological factor. About this time Professor D. Keilin started observations on a tree-hole near his laboratory in Cambridge. This had a pH of 9·5, a value on the alkaline side which few ordinary freshwater habitats exceed. Over a period of ten years Keilin (1927) found that all three species of tree-hole-breeding mosquito flourished in it. But there was no timely text-book to bring this piece of work to general notice.

An alkaline pH usually indicates a fairly high concentration of calcium, but the alkalinity may be due to other substances such as magnesium. An acid pH indicates a low concentration of these substances. But, in water where there is very little in solution, pH may vary considerably according to whether the plants are producing oxygen, which they do in daylight, or carbon dioxide, which they do at low light intensities. pH, therefore, gives only a rough indication of conditions obtaining in a body of water and its importance in ecology is not as great as was formerly thought.

The last chemical factor to be discussed here is salinity. Dilute sea-water is inhabited by a small number of species of which some have come in from the sea and others from fresh water. Some animals with close relatives in the sea are found in concentrated sea-water, but when the concentration is much above that of the sea, or when the proportion of the common ions differs from that of the sea, the faunal affinities are with fresh water. Most marine animals are permeable to both salts and water, and the proportion of ions in their body fluids is similar to that in the sea. A decrease in permeability of the cuticle is one of the adaptations that must be made by a species colonizing more dilute water, and once a high degree of impermeability has been achieved, it may be supposed that colonization of very high salinities is easier. One family of flies (Ephydridae) has developed such an impermeable cuticle that its representatives are not only found commonly in extremely saline waters, but one species lives in pools of waste oil in the West Indies.

Of the freshwater animals that have colonized brackish water, the most studied are the mosquitoes. Professor L. C. Beadle has found the salt marsh species *Aedes detritus* in natural breeding-places with salinity ranging from 10% to 122% of that of sea-water. In the laboratory the salt concentration in the body fluid of larvae was found to change with changes in the surrounding solution, but not to the same extent; the animal can exert some control, and its body fluid concentration bears to the concentration of the medium outside the same sort of relationship as the course of the back wheel of a bicycle bears to the front wheel. But the workings of the body of *Ae. detritus* are not hampered by quite wide changes in the composition of the body fluid, and this is an important difference between it and freshwater species, which die if the salt concentration alters only slightly. Presumably it is this ability which enables *Ae. detritus* to inhabit brackish water and to live in a wide range of salinities. Other species are apparently more efficient at maintaining a higher concentration of salts in the body fluid when in a medium with a low concentration, but they cannot survive in concentrations above a relatively low limit. When in a solution of high concentration, *Ae. detritus* apparently keeps its body fluid at a lower concentration by excreting salts.

Biology is dominated by fashion to an extent that would probably surprise most laymen. As Elton has pointed out, the unexpected result of the work of Charles Darwin, an outstanding field naturalist, was to send the zoological world indoors bent on elucidating the family tree of the animal kingdom by studying comparative anatomy. Eventually there was a revolt against this study of bones and pickled specimens by a generation more interested in how the living animal worked, and between the wars the leadership passed from the hands of the morphologists into those of people who could bring into play a knowledge of physics and chemistry almost as extensive as their knowledge of zoology. Many ecologists came under this influence and in their quest to explain distribution in terms of physical and chemical factors sometimes overlooked simpler explanations that an earlier generation of naturalists would have looked for first; a good example is described presently. It would be regrettable if there were to be too great a swing in the opposite direction, but it is logical to argue that before physical and chemical factors are advanced to explain the distribution of any species some investigation should be made of its behaviour, particularly when about to lay eggs, and of its relations with competitors and with predators. Several examples of these interrelationships were mentioned in Chapter 6. The present chapter concludes with a discussion of behaviour.

The limpet, *Ancylus fluviatilis*, is among the most tolerant of molluscs. It is common in streams where the current is swift and calcium low, and has also been seen in a shallow quarry pool which must have been warm on a sunny summer's afternoon and probably deficient in oxygen later on, for it was fouled by cattle. The one factor which limits the range of this snail is a hard substratum of stone or rock. The human observer can see no reason why it should not attach itself to the flat surface of a leaf or a log – but it does not. One may suppose that it keeps moving when on a surface that is not hard, and settles down only on rock or stone. A simple experiment would establish this point, but it has not been carried out so far as the authors know. *Gammarus* can neither cling well nor swim fast and there is no evident reason why it should be more successful in streams than many another freshwater organism. That it is may be attributed to an adaptation in behaviour rather than in form.

Behaviour is particularly important at the time of egg-laying. Females of most Ephemeroptera and Plecoptera appear to shed their eggs without much regard for the fate of their offspring; they do not always distinguish between water and other shiny surfaces. Most other insect groups make some selection of a site before laying their eggs. The selection may be made by the males as was shown by Dr R. Zahner, whose study of the demoiselle dragonfly, *Agrion*, in both nymphal and adult stages, made an important contribution to ecology. The adults fly peacefully together over the land beside slow-flowing streams and rivers but over the water the males establish territories from which they drive intruders of their own sex. Each territory appears to be bounded by a pattern of vegetation and contains a resting place from which the occupant can sally to repel invaders, catch food, or secure a mate. The female, having mated, lays her eggs in her partner's territory. The nymphs are likely to die from oxygen lack in still water and from starvation in fast waters because, although they can cling tenaciously, if the current is above a certain speed they cannot move to obtain food. Neither fate overtakes most of them because their parents rarely move away from the vicinity of slowly running water and their fathers select a territory over it.

Mr A. E. Gardner records that a certain dragon-fly inserts its eggs into sedges but does not always choose plants growing in water. This may be why it is a rare species. It is obviously important that the adults of insects that are aquatic in the immature stages should react in an appropriate manner when about to lay eggs. It is less important that animals which pass their whole lives in water should do this, particularly if they do not travel far from their own birth-places. Eggs, however, are always at risk from predation and it has already been suggested that the relatively large spherical cocoons in which the eggs of leeches and flatworms are laid, the gelatinous masses that contain the eggs of snails, and the brood-pouches in which *Gammarus* and *Asellus* carry their eggs till they hatch, are all adaptations to minimise this danger.

Many fish resorb their eggs rather than lay them if certain conditions are not available. Trout must have gravel, pike vegetation and there is a record of an American fish that will lay only in a tunnel of some kind. Those who wish to cultivate it in ponds throw in lengths of tile pipe.

The mosquitoes remain the most fertile source of illustration of the importance of oviposition behaviour on the distribution of larvae, probably because their relation to disease has led to much study. *Anopheles culicifacies*, a carrier of malaria in west and south India, breeds in ricefields when the rice is small, but disappears when the rice plants reach a height of about a foot. Detailed chemical and other investigations failed to reveal any change which would explain why. Eventually two workers managed to observe the female laying her eggs, a difficult feat since mosquitoes oviposit at night; they found that this species, unlike all the others investigated, which alight on the water to lay their eggs, flew a tortuous course over the breeding-place and dropped her eggs from the air. Any obstruction rising out of the water, such as rice plants, makes this flight impossible and so prevents egg-laying. The observers were able to bring the appearance of eggs in an experimental pond to an end by inserting all over it glass rods which projected about a foot above the water surface.

A less striking, though instructive, example may be drawn from mosquitoes at home. Two workers found that six ponds which they were investigating could be divided into two groups, according to the mosquito fauna. Detailed chemical analyses and an exhaustive study of the algae were made, and the data compiled are most valuable, because small ponds have been neglected by freshwater biologists. But the workers did not find what they were looking for – some chemical or botanical factor which would correlate with the ecology of the mosquitoes. Quite incidentally, in one of their papers, they mention that the ponds of one group were perennial, and those of the other group dried up during the summer, but they attributed no particular significance to this phenomenon. It was actually the clue they were seeking, for the permanent ponds were inhabited by species which lay their eggs on standing water, and the temporary ponds by species which oviposit on damp ground.

Of the British water-boatmen, *Notonecta maculata* glues its eggs to the surface of some solid object such as a log, a stone, or a concrete wall, and the other species insert theirs in plant tissues. Probably this is why *N. maculata* abounds in concrete tanks and similar places and is often the only species present, but is less numerous in weedy ponds where the other species predominate.

Captain Diver, whose survey of South Haven Peninsula between Bournemouth and Studland in Dorset would have been a model for the study of small nature reserves if he had lived to publish it, has observed that adults of different species of dragon-flies and hover-flies, all of which are aquatic in their immature stages, haunt different types of landscape. One will hawk up and down woodland glades, another in more open conditions where, however, there are occasional islands of trees or shrubs, and a third over an even sward of vegetation. If suitable conditions are not present, adults will not be present, and the result may be that larvae are absent from suitable habitats. This is a good example showing the importance of observing exactly what animals do in their wild state, a branch of study rather fallen into disfavour among professional scientists, and among naturalists as well, since the days of Fabre, Réaumur, and Swammerdam, though the ornithologists should be exempted from this generalization. It is a field in which the lead may come from amateurs not professionals; the authors do not subscribe to the view that amateurs can no longer contribute to knowledge because apparatus costing more than they can afford is necessary.

9. Food Chains and Productivity

Previous chapters have been concerned with different kinds of animals. This one is devoted to quantity, a study of importance at the present time when the world is beginning to realize that man is one of the few animals which at present lacks a mechanism to prevent its population reaching catastrophic dimensions. This chapter starts with an old-fashioned account of the various ways in which freshwater animals feed. It then switches abruptly from ancient to modern, and gives some account of calculations of production, the currently fashionable trend in ecology. Finally there is another change, this time from theory to practice, and an attempt is made to show how various theoretical findings can help the reader who wants to produce more fish.

First it is necessary to describe the food-chains in rather more detail than has been done in previous pages. Of the surface-dwellers, both the bugs and the beetles live almost entirely on dead or disabled land animals which fall on to the surface of the water. The whirligig beetles, gyrating upon the surface of the water, are apparently seeking food. If some land creature falls upon the surface of the water and lies there struggling, a quiescent group of whirligigs will suddenly start to move, and soon one of them will find it. The successful individual will continue to whirl around while devouring the morsel as quickly as possible. One is tempted to suppose that it does this in order that the others shall not suspect that it has found the object which has excited them all. The surface-dwellers are very rarely found in the stomachs of fish and there is nothing in the water which preys on them to any extent. In Windermere in August trout are feeding largely on land animals which have come to grief on the surface, so that they are to some extent in competition with the surface-dwellers. These also enter into the general scheme when they die and sink to the bottom of the water but, as the surface is so small compared with the volume of even quite a small pond, their contribution is of little significance.

In the open water the primary producers are the algae of the phytoplankton. They can utilize the energy emanating from the sun to build up substances of the sugar and starch type from carbon dioxide and water. Given phosphorus, nitrogen, and some other substances, they can also build up proteins. The animals cannot make use of solar energy in this way and are dependent on plants for the material out of which they build their bodies, reproduce, and obtain the energy used up by the activities of life.

The exploitation of this crop of phytoplankton is meagre because so many of the algae are too big for most of the animals, whose filter-feeding mechanism is designed to collect and ingest fine particles. One species of rotifer can devour large algae, and Dr H. M. Canter has recently observed that certain protozoans also feed on them.

Most of the planktonic animals take in small algae, bacteria, and the particles that arise from the disintegration of bodies, both those that originated within and without the lake, but it is still not known what the main source of food is. When supply is plentiful a large amount of what is taken in is not assimilated and some of the algae are still alive after passing through the animal's gut. The long-standing controversy about whether or not they can make use of dissolved organic matter is still unsettled.

Some representatives of each group are carnivorous and the phantom larva, *Chaoborus*, also feeds on other members of the plankton community.

It has been suggested already that the planktonic existence represents specialization to rigorous conditions, because an inhabitant of the open water has no protection against predators. They have, therefore, had to remain small and evolve a life history in which rapid reproduction while conditions are favourable alternates with a resting stage in the mud at other times. This means that the crop of zooplankton also is not exploited fully.

Zooplankton forms the main diet of the char in Windermere. In some lakes trout of all ages also make plankton their staple diet, though this is exceptional. Very young perch feed on the plankton until they reach a length of about six inches. Minnows also exploit this source of food, but, as they do not move far away from the edge of the lake, the inroads which they make upon it cannot be very great.

When planktonic organisms die they undergo at least some decay in the water, and give rise to both particulate and dissolved organic matter. It is probable that bacteria are the most important agents in this process, though the breakdown of the hard parts of Crustacea is brought about mainly by fungi; but the exploration of this field has only just started.

Much of the dead plankton falls to the bottom, where part is decomposed by bacterial action and part is devoured by the animals which dwell there. The bivalve molluscs have rows of minute lashing threadlike processes known as cilia on their gills, which are very large. These cilia keep a current of water passing over the gills, and fine particles in suspension are trapped and conveyed to the mouth in a stream of mucus. The worms probably feed like their more familiar terrestrial relatives by ingesting mud and digesting out of it anything which is digestible. No general statement can be made about the Chironomid larvae, for there are so many different kinds, but one feeding technique has been observed both in species which make tunnels in plants and in species which live in their own tubes. The larva spins a net over the exit to the tube or tunnel. It then undulates its body and causes a current of water to flow through the net. Particles in suspension are strained out, and periodically the larva turns round and devours net and all. It respins its trap and the process starts again.

The bottom fauna as a whole feeds mainly on detritus but some of the forms are carnivorous and feed on other members of the community. The larva of *Sialis*, the alder-fly, is one example, and others are to be found among the worms and the Chironomids.

Some coarse fish grub readily on the bottom but in Windermere the bottom fauna is not, apparently, preyed on extensively by fish. Eel stomachs contain a fair proportion of *Pisidium*, and perch stomachs frequently contain larvae of *Sialis* and nymphs of *Ephemera*, the may-fly, both of which are probably obtained mainly in shallow water.

There is another food-chain in the shallow water and the basis of it is the algal felt that grows on substrata of all kinds, including the higher plants, which are themselves not often attacked as long as they are in a healthy condition. The snails and some of the Ephemeroptera scrape algae off the substratum, in doing which they must take in a fair amount of debris and also a number of small animals. Water circulation will bring plankton within reach of some of the animals living in shallow water on the bottom. It is not known how important debris is. Much must be washed or blown in from the land and some matter produced in the lake no doubt settles in shallow water. Dr H. B. N. Hynes (1970) produces evidence that dead leaves and other debris washed in from the land is the basis of the food chain in running water. The freshwater shrimp, *Gammarus*, and the water hog-louse, *Asellus*, feed on debris.

Organic matter which has been reduced to a fine state of division is eaten by the Corixidae or lesser water-boatmen; they sweep up particles off the bottom with mouthparts which are highly modified for the purpose, and not like the piercing and sucking mouthparts found in the rest of the order. Particulate matter is the food of three other kinds of animal, all of which secure it in different ways. Sponges maintain a current of water through their bodies by means of flagella; *Simulium* larvae attach themselves in a fast current and strain the water with rake-like mouthparts; and various caddis larvae also living in a current build a special net to filter the water. The detritus feeders may be subsisting to a considerable extent on the remains of terrestrial plants and animals, and are then part of a food-chain which is not aquatic in origin.

Carnivores abound in this part of the aquatic world and have various means whereby they secure their prey. The water-beetles of the family Dytiscidae, and the water-boatmen *Notonecta* and *Naucoris* are active and powerful swimmers which can, if need be, overtake their prey by superior speed. The dragon-fly nymphs and the water-scorpions, on the other hand, lie concealed and wait until some suitable piece of food swims unsuspectingly within range. The lowest part of an insect's feeding appendages, the labium, is usually a flat, relatively immovable plate well provided with sensory hairs but otherwise not modified to take any active part in the capture or mastication of food. In the dragon-fly nymph the labium is hinged and provided with two movable teeth, and can be projected some distance in front of the insect's head to seize any passing prey with the two teeth. If the intended victim is just out of reach of the fully extended labium, the large dragon-fly nymphs can make a short but rapid dart after it by ejecting water from a special contractile chamber in the hind part of the gut. The forelegs of a water-scorpion have an outer segment which fits into a groove in the inner segment like the blade of a pocket-knife folding into the handle, and with these limbs they can seize and hold in a powerful embrace any animal which comes within reach. Like nearly all bugs they have sucking mouthparts, and once the prey is secured by the forelegs it is impaled on the mouthparts and sucked dry.

The tiny freshwater polyp, *Hydra*, hangs motionless from the leaf of a plant or other point of vantage, with long tentacles trailing in the water. Any small organism which blunders against these tentacles is harpooned by minute pointed and poisoned threads, which are shot out from special cells in the tentacles. Once secured and paralysed it is conveyed to the mouth by contractions of the tentacles and pushed into the inside of the body to be digested.

The leeches are also carnivorous. Most feed on small animals, molluscs being a favourite article of prey; a few attack fish, and one is said to obtain blood from water-birds by entering the mouth or nostrils and fixing itself to the back of the throat. The medicinal leech, the only one which will attack man, is now very rare in the British Isles.

The flatworms feed on small invertebrates, particularly worms, though as already mentioned, the diet of each species differs in conditions of plenty. Flatworms can trap animals larger than themselves in sticky threads of mucus which they produce. Once immobilized the prey is set upon by a pack of the flatworms and digested internally, but how often they feed in this way is not known. The method cannot be used frequently or the number of certain animals would probably decrease more rapidly than it does when flatworms become abundant. In the immense literature on flatworms there seem to be few records of straightforward observation. A damaged animal, exuding body fluids, is attacked at once by flatworms, but a healthy one is ignored most of the time. Why is it occasionally attacked? Flatworms are probably hungry all the time, so perhaps the victim unwittingly draws attention to itself now and then.

There is no true herbivore among the British fish, and all of them, except the char and probably *Coregonus*, feed largely on the animals mentioned above. The pike preys on them for a short period early in life before taking to an exclusive diet of fish. Trout and perch become mainly piscivorous when they are large. Trout find much of their food at the surface during that part of the year when insects are flying about and coming to grief there. Perch do not do this but, unlike the trout, find some of their food in the mud. The two, therefore, are not sharing completely the resources of the environment. A third numerous species, the eel, is also eating the same food but preys on snails more heavily, which possibly indicates that it glides in and out of the thick forests of pondweeds in a way that the pure swimmers are reluctant to copy.

The last aquatic link in the food-chain is provided by predaceous fish, especially the pike. In Windermere the pike feeds
mainly on perch in summer and on char in winter, these two
species being in the shallow water for spawning purposes at
these different seasons. Man, several kinds of bird, and otters
are terrestrial animals at the end of the food-chain. Parasites
also have a place, but, although they may be important, very
little is known about them.

The term 'food-web' is generally preferred to 'food-chain'
today, because the diet of almost every species is varied, and
the diagrammatic representation of the feeding relationships of
the species in a community looks something like a web and not
at all like a chain. Carnivores particularly take what they can,
and their diet changes as other members of the community
wax and wane in numbers. It also changes according to conditions and according to what, as far as the human eye can see,
which is not far, is no more than a whim. The food of trout in
a Lake District tarn discussed on p. 130 illustrates this.

Another big change in diet is that which accompanies
change in size. Obviously a creature such as a dragon-fly
nymph, when it first emerges from the egg, cannot tackle anything much larger than a protozoan, but by the time it has
reached its full size it is one of the most formidable of all the
smaller aquatic animals. The result is that food-webs, if accurately compiled, become so complicated that they convey little
information to anyone looking at them, which is what they are
intended to do. Nor, unfortunately, is it possible to discover
how much mortality any given predator causes.

The more complex the community the more stable it tends
to be, because if any particular food is in short supply there is
an alternative. In small pieces of water with simpler communities the balance is more easily upset. Instances are periodically reported in the newspapers of ponds where a solitary
pike, after having apparently eaten everything eatable, has become an item of news by attacking in desperation the nose of a
cow desiring to drink, or a swan, or even the feet of a pad-
dling urchin. Another extreme example, but one which is
apparently of regular occurrence, is described by Pennington
(1941), who studied the sequence of events in tubs filled with
rainwater. Algae were the first colonists and then came rotifers
which fed on the algae. They became very numerous and
reduced the algal population to a level where it no longer produced enough oxygen for the teeming animal population. All

the oxygen in solution was used up and all the organisms died. Then the cycle started all over again.

We conclude this, the natural history section of the chapter, by reverting to Windermere and stressing the independence of the two food-chains, for it is of fundamental importance. In the open water it runs through phytoplankton to zooplankton to char but a considerable proportion of the crop of the first two is not utilized by the next stage. In the shallower water the primary producers are attacked, and the animals devouring them are more diverse. Three main species of fish prey on this community and it is likely, though not proved, that a higher proportion of the primary production is eventually converted to fish flesh. Pike prey on the fish of both chains.

The amount of living matter produced in different natural waters may now be examined. The algae of the plankton can be counted and the weight of many of the common species is now known. Total weight obviously gives a more reliable measure of biomass than number, because the largest is many times bigger than the smallest. However, it is difficult to discover how much a population of the small algae at least has gained, and how much it has lost, between two samplings. For this reason measurement of activity has been preferred to counts for a long time. G. G. Vinberg, in a book that has been translated from the Russian (1960 AEC-tr-5692, 2 vols.) has reviewed this field. An older method was to enclose a sample of lake water in two bottles, one of which was covered with something that kept all light out. Both were then suspended in the lake at the depth from which the samples had been taken. Plants can use the sun's energy to combine carbon dioxide and water to form sugar: $6CO_2 + 6H_2O = C_6H_{12}O_6 + 6O_2$. The reverse process takes place when an organism respires. When the concentration of oxygen in the bottles is measured at the beginning and end of the exposure, the increase in the light bottle gives a figure from which the production of sugar can be calculated. The decrease in the dark bottle shows how much sugar has been used up in respiration and the assumption is made that the amount of respiration in the two bottles is the same. The more modern and more sensitive method is to add to each bottle a salt containing radio-active carbon (C^{14}). At the end of the allotted time the algae in the bottles are filtered and a Geiger counter will then reveal how much of the carbon has been built into the bodies of the plants.

The study of secondary production, that is the production of animal flesh, is more difficult. The fish farmer who releases a known weight and a known number of fish into his ponds in spring and in due course drains the pond and takes them all out can calculate production with considerable accuracy. He knows exactly what he put in and what he took out. The student of a natural piece of water does not; the data he must have, that is the population at any given time, the annual addition to it as a result of breeding, and the average loss can be obtained only by long and laborious efforts. One person who has undertaken this task is K. R. Allen whose study of the Horokiwi River in New Zealand (Allen, 1952) is still widely quoted (e.g. Hynes, 1970).

Allen captured large samples in nets, marked each individual, released it and then counted the number of marked fish in subsequent hauls. He also closed certain reaches and caught and counted every fish in each one. These considerable labours put him in possession of an unusually exact knowledge of the number of fish present. He knew how many were female and how many eggs a specimen of a certain size might be expected to lay. The counting of the redds and the examination of some of them to find how many eggs lay inside gave a cross check of the total number laid. It is convenient to consider the fate of one thousand eggs, a round and manageable figure, and this is the starting point in Table 6. The original units have been retained, because conversion to grammes would have made the arithmetic of the table less easy to follow. Nearly all the eggs hatched. The weight of a fish on hatching was only 1/250 oz. and therefore the weight of all the fish amounted to no more than a quarter of a pound. Six months later there were fifteen survivors, each weighing 2 ounces, which gives a total weight of living fish of 30 ounces, or 1¾ lb. to the nearest convenient fraction. The weight of the 985 fish that had died amounted to 3¾ lb. Another six months and numbers have halved again. Those fish that have not survived have added 1¾ lb. to bring the total contribution of those that have died to 5½ lb. and the seven living fish, now six ounces each, weigh 2½ lb. Thereafter the total weight of living fish falls, a point of importance to anglers, but the emphasis here is laid on the 3¾ lb. produced by the fish that died during the first six months, for it is a substantial fraction of the total production of 10 lb. Actually, as E. D. le Cren has pointed out, it is probably too high. Allen assumed that growth of all fish proceeded regularly between samples and he obtained his figure of weight of fish that had died by assuming that mortality was also spread evenly over each six-month period. This is likely to be a fair assumption for all except the first period, because it is known that mortality is heavy in the early days, particularly immediately after the yolk-sac is finished and the young must find their own food. However, even though Allen's figure that over one third of the total amount of fish flesh is produced by the very young and small fish is too high, the contribution is likely to be large.

Table 6. Production of *Salmo trutta* in the Horokiwi Stream (Allen, 1952)

The original units, ounces (oz.) and pounds (lb.), have been retained because the figures are approximations to the nearest whole number or major fraction. (16 oz. = 1 lb. = 0·453 kg)

Age	0	6	12	18	24	30	months
Number	1,000	15	7	4	2	1	fish
Weight of a single fish	1/250	2	6	9	12	16	oz.
Weight of all living fish	¼	1¾	2½	2¼	1½	1	lb.
Weight of all fish that have died	0	3¾	5½	7	8¼	9	lb.
Total weight of fish produced	¼	5½	8	9¼	9¾	10	lb.

The trout is a particularly suitable animal for a study of this kind. It is large, it lays large eggs whose numbers can be ascertained with accuracy, the eggs hatch within a short period, there is only one generation a year, and there was little movement up or downstream in the Horokiwi. It is not difficult to understand why data comparable to that for the trout are scarce. The eggs of invertebrates are very small by comparison, and few can be identified. It is known that some hatch over a long period, even though the oviposition time is short, which means that in any sample of the young that does not include the eggs, and samples rarely do, it is impossible to know how much loss any reduction in numbers really represents; the gap may have been partly filled by larvae from eggs unhatched at the time of the first sampling.

A similar complication makes calculation of production by planktonic Cladocera difficult, because reproduction is fast and continuous, and sexual maturity is reached after a short period. It is extremely difficult to culture the animals from the open water of a big lake in captivity in conditions that might make information about frequency of reproduction, number of young per birth, and age of young at sexual maturity applicable to a wild population.

Up to this point production has been discussed in terms of weight. It is more accurate to discuss it in terms of energy. This sometimes appears to be an unnecessary complication to the layman, but the housewife would have no difficulty in explaining one good reason for doing it. At the butcher's she knows that a pound of meat may be a pound of edible flesh or may include a weight of bone which, even though the dog may enjoy it, consists largely of calcium salts, which the body requires in small amounts only. 'Ecological energetics', as the current jargon has it, has been much studied in recent years, and in a book of that title, J. Phillipson (1966) has set out to describe it clearly, simply, and shortly.

Production, as will have been evident from the foregoing, is used here to mean the production of living matter in a limited period of time. Within a year the population of a large mammal will have done no more than produce enough offspring to make good relatively small losses; a pond-snail, which dies after laying eggs, will have produced the whole population except that portion which the preceding generation contributed to the egg; smaller animals may have reproduced themselves many times over. The element of time in the definition is important because in the long run loss balances gain and the energy that runs through the living organisms of the world keeps the system working without loss or gain. This must obviously be so or the whole world would have been overpopulated or depopulated years ago.

Energy in the form of heat from the sun pours onto the earth at the rate of about 15.3×10^8 calories per square metre per year. Plants can utilize this energy to make sugar and proteins. In other words they are converting the free or kinetic energy into potential or stored energy. This transformation is constantly proceeding in both directions, but the amount of energy involved never alters. The potential energy in the plants is the source from which animals directly, or indirectly if they are carnivores, derive the energy which enables them to move and r̄ ˜orm all the functions of life. It has been compared to the petrol that goes into the tank of a car or the coal that goes into the furnace of a steam engine, but the analogy cannot be taken too far because living organisms use the same material to repair themselves, to grow and to reproduce.

The amount of energy in a piece of tissue from a living organism is ascertained by burning it and measuring the amount of heat produced. When it is burnt inside an animal the measurement of heat is less easy, but the amount of oxygen used can be discovered easily. Any chemical reaction always yields the same amount of energy, shared between work done and heat generated, and the release of energy in nearly all living organisms depends ultimately on oxidation.

One calorie is the amount of heat required to raise the temperature of one gramme of water one degree centigrade. One calorie is the equivalent of 4·2 joules. One joule is the equivalent of ten million ergs, of which 981 are required to raise one gram one centimetre against the force of gravity.

It is convenient to take as an example of an exercise in this type of ecology a spring studied by J. M. Teal (1957) in America, for it is also used as an example by Phillipson. If the present account lays more stress on the shortcomings of the study than Phillipson's does, there is no intention on the part of the authors to disparage the work; the only writer who can safely criticise the findings of a pioneer is one who has no ambition ever to be a pioneer himself. A presentation that is fair to the reader must draw attention to gaps that at present are bridged by assumptions, and Teal himself writes: 'It should be emphasized, however, that in the present state of our knowledge of community metabolism considerably more assumptions have to be made in order to present a complete picture than would be the case in many other fields.'

The spring was convenient for several reasons. It was small and had a restricted fauna. Over forty species of animal were recorded but only nine were sufficiently numerous to justify investigation. Physical and chemical conditions remained nearly constant throughout the year and no animals could be washed down from above. Migration upstream appeared to be negligible.

A chironomid, that is a non-biting midge, was comparatively easy to study because it spent the winter in the egg stage, and all the eggs were assumed to hatch within a short period, though no evidence that they did is brought forward. Emerging adults were trapped and the total number emerging from the whole spring calculated from the captures. Observation indicated that one third of the females flew elsewhere, and the rest laid eggs in the spring from which they had come. Dissection showed an average of 250 eggs per female, and there was therefore all the information necessary for a calculation of the number of eggs laid. It was found to be 980,000. It is stated that 87% did not survive to hatch, which is not improbable if they lay in the mud for seven months exposed to predation, but this calculation could be subject to a big error. First instar larvae were small enough to pass through the sieve and therefore they were not counted. The calculation of their number was based on the assumption that mortality was constant between each instar. It could be assumed with equal justification from the information published that survival and hatching of eggs was good and most of the 87% mortality occurred early in larval life, for this was the pattern found by Allen studying trout.

The number of larvae per square metre was found by pushing a sampler of known area into the mud where the larvae lived, sieving what was brought up, and picking out the larvae. The results are shown in the first line of figures in Table 7. The figures for the first month meant little because, as already explained, first instar larvae passed through the sieve. Weight was converted to kilocalories by means of a chemical process, not by combustion in a bomb calorimeter which, as described above, is the usual method today. Mortality presented no difficulty and, once all the eggs had hatched, which was believed to take place within a short period, it was shown by the fall in numbers from one sampling to the next. From this information the production, shown here in the upper part of Table 7, could be calculated. The total was 130·7 calories of which nearly three-quarters was contributed by individuals that died during development, nearly one quarter by those that survived to adulthood, and the rest by the skins shed at the moults.

The lower half of Table 7 shows the amount of energy used during the aquatic stage as revealed by the amount of oxygen used. A figure for this was obtained by incarcerating larvae in a container which was returned to the spring for a period sufficient for the reduction of oxygen to a level easily measurable by the analyst but not limiting to the larvae. 'The water was usually kept sufficiently stirred by the activities of the animals themselves' writes Teal in a passage which must excite some speculation about the meaning of the results. Wautier and Pattee investigated the consumption of oxygen by nymphs of *Ephemera danica* at different temperatures. The nymphs were inside a glass flask and the oxygen consumption varied enormously according to whether a substratum was provided and according to the nature of the substratum. If the flask was floored with sand, the substratum which the nymphs normally inhabit, oxygen consumption rose only slightly as the temperature was raised. If the flask contained nothing, consumption rose rapidly as the animal tried to escape from its unnatural surroundings. A substratum of pebbles produced an intermedi-

Table 7. Population, production and consumption by the chironomid, *Calopsectra dives*, in Root Spring (Teal, 1957).

Population							
	Jan. April	May	June	July	Aug.	Sept.	Oct. Nov.
Larvae							
number/m^2	0	1,700	89,500	65,000	57,000	200	0
weight in g/m^2	—	3·0	58·5	82·4	127·0	0·2	—
k cal/m^2	—	2·1	40·4	56·8	87·6	0·1	—
Adults							
number/m^2	0	13	170	953	3,464	13,250	533
weight mg/m^2	—	12	156	876	3,170	12,200	490
k cal/m^2	—	0·014	0·246	1·38	5·00	10·3	0·775

Production	
adults	26·7
pupal skins	11·8
larval skins	18·5
larvae and pupae that died	73·7
	130·7 k cal/m^2/year

Consumption	
respiration by living animals	312·6
respiration by animals that died between measurements	61·6
To deposit	15·4
	389·6 k cal/m^2/year

ate result. It seems reasonable to suspect that the larvae enclosed by Teal without their familiar substratum were unusually active and that his figure for respiration is too high.

The total intake by the chironomids is 520·3 kilocalories per square metre in a year. Of this 389·6 kilocalories were used during the course of the activities of life, and 130·7 was devoted to the production of tissue. This is 25%.

Teal gives no information about what fraction of the food eaten by the chironomid larvae is utilized, but he did obtain a figure for a carnivore. It was fed prey of known calorific value in such quantity that it devoured all of it. Assimilation, that is the amount taken into the tissues, was discovered from amount of weight put on and amount oxidized. The difference between this and the value of the prey eaten was the amount wasted, that is passed through the digestive tract unused.

Two other animals investigated by Teal, *Limnodrilus* and *Asellus*, have been selected for discussion here because, unlike the chironomid, they are present all the year round. Certain facts about the worm, *Limnodrilus*, are presented: the numbers per square metre each month are shown in Table 8; it reproduces all the year round; full guts are found through the year; the caloric content was 0.76 ± 0.026 calories per milligramme of fresh weight; the maximum rate of increase of an animal kept in the laboratory in mud from the spring was 0.474 calories in 30 days; 1.22 ± 0.35 calories per calorie of fresh weight were oxidized per month.

Numbers collected today subtracted from numbers collected last time gives mortality, provided there has been no addition to the population between the two samplings. If there has, biomass, which is numbers multiplied by average weight, times rate of growth will give a figure for any future date, and the amount by which the actual figure obtained falls short gives mortality. How to measure rate of growth and reproduction; how did Teal arrive at the figure of 0.474 quoted above? It is clearly a crucial question. The fairest step is to quote the author. 'Weighed animals confined in strained mud in the laboratory cold room were reweighed after two weeks. When one or more animals in an experiment died, the experiment was discarded. Of the five successful experiments the one with the maximum rate of increase, in which k equalled 0.474 for a thirty-day month, was taken as the significant one since laboratory conditions were not as conducive to growth as were conditions in the spring to which this population of worms was adapted.

A value for the rate of increase was also obtained from the increase of the natural population in the spring from April to May, $k=0.530$. Even though the P value [weight of population] lay between 0.05 and 0.10 for this population increase, the larger value for k was used as it agreed fairly well with the maximum value obtained in the laboratory experiments and as

k obtained from the fluctuations of a natural population will have a minimum value.'

This is the sort of figure which physicists seize upon to deride biologists as practitioners of an inexact science. Moreover the calculations involve certain assumptions. The first is that this figure is applicable to the whole year regardless of seasonal variations and changes in population. The temperature of the spring did range over 4° C. between winter and summer and the changes in the length of day affected this biotope as much as any other. It has been shown for Pacific Salmon that numbers can depress growth before any shortage of food comes into play and this could be true of worms also. It is also assumed that the same figure embraces growth of each individual and the production of young.

A drop in population between October and November 1954 is attributed to hurricanes. The other changes are not explained, though some naturalists may feel that such large fluctuations in a uniform environment by a species that is reproducing all the time ought to be explained before any further steps are taken. One of the outstanding difficulties of the study of animal populations is due to their ability to move, to disperse from the place where the eggs were laid and often to congregate in small areas of which the attraction is not perceptible to the human eye.

Evidently the final results are liable to an error whose size is unknown but which may be great. It is the naturalist who likes to base the advance of knowledge upon exact information about the way of life of each species who snipes at the study of the whole. Fire comes from the other side, from the physiologists, at the work on the chironomid. Well may Phillipson write; 'Without unlimited numbers of research workers a thermodynamic study of any living thing must be restricted.' However, the work leads to results of intrinsic interest and considerable practical importance and if the pioneers do little more than stimulate others to produce more accurate figures their achievement will have been no mean one.

The respiration of all organisms was obtained from the fall in oxygen concentration in a given time in a sample of spring water and mud. Subtraction from this of the sum of the figures for individual species gave the respiration for the microscopic organisms, which proved to be 350 kilocalories per square metre per year.

The net efficiency, that is amount produced as a percentage

of amount assimilated, was about 20 for all the herbivores and detritus feeders except *Pisidium*, which achieved 47%. It was higher in the carnivores, 59%, and planarians alone reached 87%. However, if the mucus which they produce is treated as an expenditure of energy, the efficiency is found to be only about 30%.

The figure for a chironomid which is partly carnivorous must be reduced because it manages to consume only about one third of each item of prey. A small carnivore devouring piecemeal an animal not considerably smaller than itself, will lose much of the body fluids at least; a larger one capable of swallowing the prey whole must be more efficient.

Words such as 'interesting' or 'important' are generally no more than matters of opinion in academic research, and it is easier to conclude this account of energy flow from a practical point of view. The sport fisherman whose prey grows in the wild is concerned primarily with the efficiency of each step in the food chain, in other words he wants to know how much of the biomass at one feeding level becomes biomass at the level above. Teal summarizes his findings in a diagram that has been widely quoted. It starts with an input of debris and algae. The debris consists of dead leaves of which the quantity falling into the spring was discovered simply by means of trays of known area. The production of algae was measured by the light-and-dark-bottle method. In the middle of the figure are two squares, one for herbivores and one for carnivores, the area of each being roughly proportional to biomass. A figure shows the difference between biomass at the beginning and end of a season. Actually it is −4 in both boxes, but obviously over a longer period it would be nil. There is no overall gain or loss, but energy flows through a system of a given size and is used in maintaining that size, though within a limited period the size may vary greatly. It is a simplification to refer to herbivores for earlier in the paper Teal writes: 'The most abundant animals were those which fed on debris and algae, taking mud into their gut and assimilating the digestible material.' It seems improbable that they did not digest animal remains in the debris and omnivores seems a preferable word.

To come to the figures, algae contributed only 655 kilocalories after respiration had been taken into account, and the main primary source of energy was the debris, 2350 kilocalories. Of this total of 3005 kilocalories 705 were not used and 2300 were assimilated by herbivores. Of these 1746 were used in respiration and converted to heat, leaving 554 (and 18 kilocalories from immigration) to pass on. Of these 337 went to debris and 208 were assimilated by carnivores, and 31 left the system as adult insects. Contributions to deposit were 705 by the debris, 337 by the herbivores and 121 by the carnivores, and of this total of 1163, 295 were utilized by microorganisms. Phillipson (1966, p. 27) opines that much of this debris is washed out, but no figures for less are quoted. The point of querying the term herbivore above was to justify the suggestion at this point that some of the dead bodies in the debris were used by the herbivores and that an unknown portion of the 868 kilocalories in the deposit should be added to the figure for debris coming in.

The figures to be picked out are the 2300 kilocalories assimilated by herbivores and the 208, roughly 9%, assimilated by carnivores. Laboratory experiments carried out by L. B. Slobodkin, and also described by Phillipson, had indicated that each successive link in a food chain would be about 10% of the one before. The angler in whose pond efficiency reaches that level will be doing well. For the moment let us leave that convenient round figure to return to later on.

The production of food rather than the provision of sport involves consideration of some of the other figures which Phillipson has brought together. For example a certain mussel respires only twice as much of what it takes in as it uses for growth. Two fish, roach and bleak, use 13 times more energy for respiration than for growth, and the figures for some other animals are: weasel 42 times; field-mouse 55 times; and Savannah sparrow 89 times. These are figures for a year. Time is an important consideration in production. From a ton of hay one may produce 240 lb. of beef or 240 lb. of rabbit meat, but the beef will take four months to grow to that amount, the rabbits one. The above figures bring out the point that carnivores are more efficient than herbivores. Digestion of plant material is less easy than digestion of flesh, and many herbivores fail to assimilate a considerable proportion of what they eat. Another point that comes out is that cold-blooded animals are more efficient than warm-blooded ones; the latter may have to devote energy to keeping warm and all the bodily functions operate at a constant rate regardless of the outside temperature. Those of the cold-blooded animal often slow down as temperature falls, which makes it more efficient at converting its food to flesh though not more efficient in the struggle for existence. One imagines that a cold-blooded animal being chased by a warm-blooded animal in a rapidly falling temperature feels like a man in that familiar nightmare on the railway line with the train bearing down and the limbs afflicted with almost complete paralysis.

For food production, then, a herbivore is better than a carnivore, because under given conditions it will yield ten times more. The warmer the climate, the better the cold-blooded animal. The ideal animal grows quickly and is cultivated only during the time when it is growing fast.

By way of contrast to the generalizations and assumptions inseparable, at present, from the study of a whole community, it is instructive to examine what is known about an animal that has been thoroughly investigated. This brings us back to the trout which has been studied in detail in the laboratory by Dr M. E. Brown. An animal needs a certain proportion of its food to provide the energy for the activities of life and for repairs. Brown refers to this as the maintenance requirement. Until it has been met, food is not available for growth. At a temperature of 11·5° C. a fish weighing 100 g. requires 6.5 g. per week for maintenance. At 7° C. the figure is about 5 g., and at 15° C. about 13 g., and the fall and rise beyond those temperatures is slight. A smaller fish has a larger requirement and a larger fish a smaller one. These figures were obtained by fish living in an aquarium and fed regularly, so that they used little energy searching for prey and pursuing it, or in fleeing from enemies. The consumption of oxygen during these activities can be estimated only roughly by anyone attempting to calculate the energy flow in a wild population.

Growth was not constant under constant conditions. Two-year-old trout kept at 11.5° C. and twelve hours of light per day grew fastest in February, after a minimum in October and November, and then progressively more slowly till the autumn when they became sexually mature.

When young trout were kept together they soon became separated into groups according to the rate of growth, even though there was no shortage of food. If the small and medium-sized fish were removed to a tank on their own they grew well and were soon similar in size to the fish which had always been large. The growth of two-year-olds did not seem to be affected by the volume of water but it was affected by the number of companions, of which both too many and too few resulted in slower growth.

Food was used most efficiently just above the maintenance level. Frost and Brown (1967) write that under optimal conditions in a hatchery 1 pound of trout flesh can be obtained from 2 lb. of food through the year. However the food mentioned is dry pellets, and the calorific value of 1 lb. of them will be higher than that of 1 lb. of trout flesh which contains a high proportion of water. The same authors write that a net conversion of 25%, that is 1 lb. of flesh from 4 lb. of food, is an average value for laboratory conditions. Net conversion, it must be noted, is conversion of food after the maintenance requirement has been subtracted. Trout grow relatively slowly. The best growths quoted by Frost and Brown (1967 p. 101) are—

first year	170%
second year	96+%
third year	56%
fourth year	31%
fifth year	23%

These are percentage increases in length during a year, but calculated in a way that treats increment as if it were compound interest rather than a simple difference between length at the beginning and end of the year. Slow growth means that the maintenance requirement will always take a substantial fraction of the total food assimilated and therefore the ratio of growth to food will vary considerably according to the rate of growth. This, as has been shown, depends on size of fish, temperature, and other factors discussed above. It is not unexpected, then, to find that calculations of pounds of food required to produce one pound of fish range from 2.3 to 7·1 (Brown, 1957).

These are the ways in which various authors have approached the problem and it is appropriate now to present some of their results. Although the C^{14} method has been in use for several decades, it has been applied sparingly. The reason is not hard to find; the measurement of primary production in a lake requires frequent observations through a year, for production will vary from day to day according to the intensity of solar illumination and according to the density of algae, which can change greatly in a few days. Moreover in summer a very long day's work is necessary. On a sunny day when algae are abundant, production will be intense a metre or two below the surface, because in shallow water the light is strong enough to hinder it. Production drops abruptly below the zone where it is at a maximum, because the dense population of algae above has cut off the light. In an unproductive lake the amount of production changes much less with increasing depth and the zone in which plants are active is deeper. Results are sometimes expressed as weight of carbon produced in the cubic metre where production is greatest. This is a useful comparative figure. Professor I. Findenegg has stressed that the shape of the curve obtained when production is plotted against depth may be more useful for comparing lakes than a figure. Total production can be obtained by integrating the values obtained between the surface and the depth at which there is no production beneath one square metre. Jónasson and Mathiesen (1959) quote the following figures:

Lyngby Sø	660 g C/m²/year
Esrom	180
Erken	104
Lunzer Untersee	30

Esrom is a typical productive Baltic lake in Denmark. Lyngby is similar but shallower and more enriched by sewage. Erken is in Sweden and the Lunzer Untersee, in Austria, is an alpine lake, which, however, is likely to be some way above the bottom of any series based upon production. There are no comparable figures for any British lake and comparisons can be based on standing crops only. Dr Lund has pointed out that the numbers of the larger algae are not reduced much by grazing and that they continue to multiply until some chemical requirement in the environment is exhausted. Therefore, if the numbers of each species can be ascertained when they are at a maximum, an indication of the potentialities of the lake in question will have been obtained. The very small algae can be ignored, as their contribution to total biomass is slight; they are eaten by animals and therefore the total in any sample is not a measure of the total produced. The only published results are based on numbers, but now the weights of many more species are known and Dr Lund has kindly put at our disposal the figures for average dry weight of algae produced in the Lake District lakes in a year (Table 9, p. 191). Production in Esthwaite is seen to be some 300 times that in Wastwater. Production in Ullswater is higher than might be expected, but otherwise the order into which the lakes fall is similar to that deduced from other studies.

The collections were not made in such a way that the figures represent exactly either the production in the cubic metre in which it was greatest, or the total production in a column beneath one square metre. It is likely, therefore, that production in the most barren lakes, in which the process extends to greater depths than in the others, is relatively higher than indicated. Bassenthwaite in particular is overrated because the depth at which algae are active is reduced by the peat-stained water (Table 2, p. 72).

It is convenient to pass on to the figures for the animals, shown in the same table. They were obtained by collecting with a hand net in water sufficiently shallow to allow the collector to reach the bottom and lift up stones with his or her arm. All were on the stony substratum, which differs little from lake to lake. They were made during the winter and spring when the population is at its most stable, additions from hatching eggs and losses due to the emergence of adult insects, being slight.

On these figures Esthwaite is some 30 times more productive than Wastwater (the small population in Thirlmere is attributable to the big range over which the water rises and falls in this reservoir), compared with the 300 times mentioned above. It may be noted in passing that the contribution of Plecoptera to the total fauna tends to be greatest in the least productive lakes and no species in this order has more than one generation in a year. In lakes near the middle of the series Ephemeroptera tend to be more important and some of them have two generations a year. This is true of *Centroptilum* but a species that contributes more to the difference between the lakes is *Ecdyonurus dispar*. This is numerous all through the year in the north basin of Windermere, where there is one generation that overwinters and another that grows more quickly in summer. In less productive lakes there is only a summer generation and no nymphs are to be found in winter. Near the upper end of the series insects are scarce and such animals as flatworms and crustacea are important. The flatworms are unique in that biomass adjusts itself to resources. They do not breed in the winter and when the temperature passes the level at which reproduction starts the population is generally less than the water could support. Young continue to appear till this is no longer true and, if food should be short, pregnant flatworms absorb the eggs maturing within them and then, should supplies still be insufficient, decrease in size (Reynoldson, 1966). *Gammarus* reproduces continuously from early January to the end of October. The flexible reproduction of the flatworms leads to a maximum exploitation of resources and *Gammarus* is not far behind them in this respect. However, if the insects with one generation a year can make good losses of large specimens from a reserve of starvelings or unhatched eggs in the way postulated above (p. 131) and later in this chapter, which is possible though it has not been demonstrated, they too can exploit the resources of their surroundings to the full. The one advantage obviously possessed by the flatworms and *Gammarus* is that they can recover more rapidly from any catastrophic reduction in numbers. Evidently the relation between standing crop and production will not be the same for each of

Table 9. Comparison of phytoplankton and animals in Lake District lakes

Lake	Average dry weight of phytoplankton produced in a year †	Number of animals collected on a stony substratum in 100 minutes
Wastwater	1,100	194
Thirlmere	2,637	48
Buttermere	3,356	455
Ennerdale	4,492	515
Crummock	5,218	847
Haweswater	17,794	—
Coniston	38,350	1,294
Windermere N. basin	68,029	2,129
Derwent Water	90,928	1,307
Loweswater	105,387	2,739
Windermere S. basin	136,596	2,815
Blelham	175,260	—
Ullswater	185,346	1,696
Bassenthwaite	205,702	2,508
Esthwaite	336,466	5,987

† Weight in mg of the number of cells in one ml × 10^6

these three different types of life history. The animals whose cycles they are occur in different proportions in the various lakes, and therefore the figures in the right-hand column of Table 9 do not provide material for exact comparison. A rough figure for the conversion of these figures to numbers per square metre can be obtained in two ways. Calculations from the results obtained in a quantitative sampler in a stream indicated that a square metre was covered in 50 minutes; the figures in Table 9 must then be divided by two. The only quantitative data for lakes are those obtained by H. P. Moon in Windermere more than thirty years before the work that produced the figures in Table 9 was done. This suggests that a square metre was covered in a shorter time and that the figure by which the totals in the table should be divided to give numbers per square metre is nearer 3 than 2.

For the present the approximation is sufficient to show that Esrom Lake is far more productive than any Lake District water. The stony shore is limited to the shallowest water in this lake, and at 2 metres the bottom is sandy. It continues sandy at greater depths with an ever-increasing amount of organic matter, and it is not until 20 m. is reached that the sediment is entirely of organic origin. In the 'surf-zone' on the stones at 0.2 m. there are some 5,000 organisms per square metre. At 2 m. on sand there are twice as many. The number is back to near 5,000 at 5 m. but up once more to 1,000 at 8 to 9 m. at which depth *Dreissena*, the zebra mussel, begins to be numerous and to cover most of the bottom. Total weight is far greater here than at any other depth though much of it is mussel shell. Thereafter numbers and weight decrease regularly to 20 m. where there are some 4,000 organisms per square metre. Berg (1938), who carried out the survey of Esrom Lake, devotes a section of his account to comparison with other lakes. Among them there are no British waters. Miss (now Professor) C. F. Humphries made some observations in Windermere and found an unexpectedly sparse fauna.

During the course of the study mentioned, K. R. Allen discovered that fish appeared to be eating about seventeen times as much as there was to eat. Other workers have experienced 'Allen's paradox' (Hynes, 1970). Obviously until this paradox has been explained, calculations of production in running water cannot mean very much. No full explanation is possible at present but certain lines of investigation which, when followed up, should provide one, are worth mentioning. First there is the difficulty of ascertaining how much a fish eats. One can discover easily enough what it contained at the moment of death, but how long would it have taken to digest that food, and would it have kept its stomach permanently as full as that? The first question has been answered by laboratory studies and, on the results, Allen and others have based their calculations. The second is much more difficult to answer. D.

R. Swift has shown that trout are more active at some times of year than at others. Moreover, if a trout that has filled its belly lies quiet fasting for a period thereafter, most methods of capture will not secure it and will bring in a sample of fish active at the time. The whole question of fasting by species that grow slowly is little explored and difficult. However, since Allen's paradox has also been revealed by calculations of what a given population of fish require and what is available, it is likely that the error is to be found in the sampling of the invertebrates, not in the calculation of what fish eat. A calculation of production from standing crop assumes a knowledge of life cycles, which is generally sufficient to obviate large error. Delayed hatching of eggs is known in several stream-inhabiting insects and, as already explained, this can introduce a big error into calculations. The observation on dragon-flies in Hodson's Tarn may be relevant in the present connection (p. 131). There, it will be remembered, the number of large nymphs was not much smaller when trout were present, and were eating nymphs, than when there were no trout. The explanation tentatively offered was that the ranks of the large ones were filled from a reserve of 'starvelings' living in places where they could obtain enough to keep alive but not enough to grow. It is possible that there are big reserves of this kind in running water. They may drift down from small tributaries too shallow for fish to penetrate; movement downstream or up is a factor that will always complicate the study of water courses. The most important source of reserves is more likely to be the gravel and finer particles, deeper than most sampling goes, for it is known that many tiny nymphs are to be found there. When these suggestions have been investigated further, and when numbers have been obtained, calculations of production will be more reliable. In the meantime it will suffice to mention that Hynes (1970) records that he himself calculated that the production of invertebrates in a Welsh mountain stream was 30 kg./ha. wet weight, roughly the same as the production of fish in an unproductive pond, and therefore unexpectedly low. He quotes a much higher figure for a stream that seems to be similar.

It is pertinent to interpose into this discussion of animal numbers some account of how they are limited. Ultimately it must obviously be by the amount of food. Also obvious is the likelihood of selection pressure in favour of some mechanism whereby the available food is shared among an élite and not consumed by a horde large enough to eat everything and bring about the starvation of the whole population. The most usual means is the establishment of territories. When young trout begin to feed, each one attempts to take up a territory on the floor of the stream and to hold it against all invaders. As the trout grow, they enlarge their territories and those unsuccessful in this struggle die. H. Kalleberg has watched this in a large glass-fronted tank in Sweden and E. D. le Cren experimented in a natural stream. He confined different numbers of young trout in stretches of equal length and found that the final population in each one was almost identical. Certain small dragon-flies take a year longer than usual to attain full size, when the population is large, and it is thought that only a limited number find and hold feeding places where they can catch enough for normal growth. Large tadpoles prevent the growth of smaller ones by a secretion.

Populations may not reach a size where natural control is necessary. Catastrophic reduction in streams with unstable bottoms may occur during floods. In standing water there is sometimes a risk of deoxygenation on hot still days in summer. Of more general importance, however, is predation. An animal must have some kind of cover in which to take refuge from predators, and the size of many a freshwater population is governed by the amount of cover available. From the predators' point of view there can be too much cover; fish will fare equally badly in a pond devoid of vegetation and one overgrown with it.

Cover reintroduces the practical aspects. Academic research, described above, has shown that from 1,000 lb. of vegetation one should hope to obtain 100 lb. of herbivores, 10 lb. of primary carnivores and 1 lb. of secondary carnivores, which is what a fish is likely to be. If this ratio is not obtained, applied science should show why, and discover how to improve it. An example of an unsatisfactory food chain was seen in the open water of Windermere where many algae are too big to be eaten by the animals, and the animals are so small that the fish which prey on them must spend a lot of energy filling their bellies. It was suggested that nothing of intermediate size has evolved because it would be too easy for predators to catch in the coverless conditions of the open water. Where cover does exist, there is probably a continual egress from it of surplus population seeking a territory and it is on this surplus that the carnivore feeds. Unless an animal can graduate to larger food as it grows larger itself, it will use up more and more energy finding its food. Frequently there are links in the food chain that are undesirable from the fish-producer's point of view. In any community there are a number of small carnivores which fish eat so that the fish may be the fourth, even the fifth link in the chain, not the third as indicated above. There are also many parasites, generally neglected in calculations of production, though possibly of considerable importance. The shell of a snail heavily infested with trematodes may appear to enclose more parasite than host tissue, and enormous production of young is the parasite's reaction to the big problem in its life-history – finding a host. Finally, the food-chain may terminate in some undesirable species; for example the rich supply of material floating down the Thames is used by a large population of large swan-mussels, which are apparently rarely eaten by any other animal. There is therefore a large biomass in which material is locked up until the individual dies and decomposes.

Clearly many of these departures from his ideal cannot be corrected by man. There are, however, certain practical steps that can be taken, and also certain unpractical steps that often are taken. The latter may be disposed of first. There is nothing easier than to add something to a piece of water. Occasionally an exotic species is added, which is generally a failure, because the creature does not succeed in finding a niche that is not occupied by a native. If the introduction is not a failure, it is often a disaster. On other occasions the addition is of some species that is not found in the water in question though it occurs in others not far off. In a newly created pond or lake or one that has been depopulated by an accident, this may be a profitable step, but in an established one its value is doubtful. Freshwater animals traverse land barriers with some ease, though how they do it remains a mystery. One of the characteristic features of the freshwater environment, the one in which it differs most from the land or the sea, is that it consists of a great many isolated pieces, most of them small. Any animal that is successful in this environment must over the years have evolved some means of getting from one to another. Therefore, the likelihood is that, if a species is absent from a pond, it is because conditions are unsuitable and not because it has never reached it. Introduction is not likely to be successful.

Sometimes large numbers, or what sounds like large numbers at a committee meeting, of a species already present are introduced to increase the population. Snails are a favourite animal for this exercise. The committee generally consists of persons unaware of the enormous numbers already present in any natural water or of the immense number of eggs produced by most invertebrates. We recall the late F. T. K. Pentelow, who, referring in an article to the coarse-fisherman's habit of returning his catch alive at the end of the day, asked plaintively 'What do they think the fish do in the breeding season?' An introduction probably augments the natural population much less than those responsible imagine, and effects no more than a temporary increase because the size of the population is determined by factors within the pond or lake.

Fly boards are pieces of wood moored in a river. Species of *Baetis* that normally crawl down some solid object fast in the bottom to lay their eggs on it under water, lay their eggs on the submerged side of the board. The eggs there are much less accessible to predators, stated to be caddis larvae which cannot swim, and many more hatch. Whether this leads to a bigger crop of moderate-sized or large larvae has not been shown scientifically, and is doubtful, for the mortality among tiny nymphs is colossal. The number which survive probably depends on the environmental conditions and is independent of the number of eggs that hatch.

Positive action can be taken at various points of the food chain. Production can be increased by adding nutrients to the water and an account of this is given in Chapter 13. At the other end species of fish deemed to be undesirable can be reduced in number and an essay along these lines is described in Chapter 12. The middle parts of the food chain can be affected by altering the amount of cover. Most of the animals on which fish feed seek cover in vegetation and the number in a given volume is probably often determined by the nature and density of the vegetation. This rarely grows in the way the angler would wish and is either too thick or too thin. In south country chalk rivers the weed has to be cut drastically in the interests of drainage. Where a keeper has a free hand he will probably have views on how best it should be done, and his empirical approach based on long experience is probably sound. So far no biologist has made the observations that would justify an opinion on the size and shape of patches of weed that would yield the maximum amount of food, and the size and shape of the channels between them that would suit the fish and the fisherman best. Recently much attention has been paid to the eradication of vegetation chemically (Robson, 1968).

In moorland ponds there is often a poor growth of vegetation, probably because, after a time, the bottom becomes unfavourable to rooted plants. The late Mr E. Langman, keeper at the Wray Mires hatchery now rented by the Freshwater Biological Association, used to throw bundles of bracken into the tarn to provide cover for invertebrates. At present experiments with artificial vegetation made of polypropylene rope are being carried out in the same tarns. I.C.I. manufactures artificial seaweed but so far only for the purpose of stabilizing sea beaches.

Up to this point the discussion has been concerned with water shallow enough for vegetation to grow everywhere. In lakes where a considerable proportion of the bottom is too deep for plants to establish themselves, the food chain is unsatisfactory for reasons already set forth – unsatisfactory that is from the point of view of the fisherman. How better use might be made of the potential of a lake was revealed by some experiments carried out by the Freshwater Biological Association for a different purpose. Part of a small lake that stratified and became deoxygenated in summer was dammed off so that the effects on the chemistry and on the algae of preventing those two events by injecting compressed air could be studied. The dam was a plastic sheet. It was noticed that the upper part of this became covered with attached algae which grew thickly enough to provide cover for snails, shrimps, and insect larvae. The nutrients entering the lake, instead of being taken up entirely by algae whose ultimate fate was to sink to the bottom or to be washed down the outflow by flood water, were being partly taken up by attached algae that provided cover and probably food for the kind of invertebrates on which fish do well. Whether this is a practical method of increasing the production of fish in a deep lake has not yet been tested.

In the upper reaches of rivers floods may keep the population of invertebrates below the theoretical maximum if the bottom is unstable. In the lower reaches silting and bank erosion may be problems. Biological problems and engineering problems mingle and the latter probably take precedence. The reader is referred to books on fishery management such as that of Fort and Brayshaw (1961).

10. Life Around the Water

A rigid distinction between water and dry land is one of the main aims of man, and his efforts in pursuit of this objective produce one of the major differences between country which he has colonized and country which he has left untouched. Many plant associations which once covered great areas of Britain are now represented by fragments, and often the animals which dwell in them have been even more affected; for an animal is influenced by the average conditions of a wide area, whereas the presence or absence of a particular plant species may depend on the conditions obtaining within a quite limited compass. Before these communities are examined some of the terms commonly applied to them may be noted, for Tansley (1939) has suggested certain precise definitions based on accurate botanical surveys. The four terms are swamp, marsh, fen, and bog. Swamp is wetter than the other three, and at no time of year can a man traverse a swamp without getting his feet wet. It is the final stage of development of an aquatic habitat. It was doubtless widespread in the fen district at one time but elsewhere was more limited than the other types of community. The typical vegetation of swamps is emergent and consists of reeds, sedges, reedmace, and other plants, according to conditions, some of which have been described on p. 118 in connection with the development of reed-beds in sheltered bays in Windermere.

In winter or at any time of year after heavy rain, marsh, fen, and bog may be well inundated, but the normal summer water level is only just above the ground level. The soil is always waterlogged, but in dry spells a man may pick his way across these types without getting his feet wet. Marsh has a mineral substratum. A glacial lake, particularly one on a flat plain, may be surrounded by extensive areas of marsh growing on the boulder clay. In earlier times vast tracts of land on either side of the lowland courses of rivers were covered by marsh, growing on a subsoil of gravel, sand, or silt deposited by the river in flood (cf. Fig. 6, 62). Alderwood is the typical plant community of marsh conditions in Britain.

Fen and bog both occur on a substratum which is mainly

organic, that is derived from the accumulation of plant remains, and the difference between them is that bog is markedly deficient in lime and other bases, and fen is not. In Britain the classic type of fenland occupies a great stretch of country in North Cambridgeshire, South Lincolnshire, and parts of the adjacent counties. All the water which flows through it comes from the chalklands or limestone and is rich in bases, but there is little oxygen in the waterlogged soil. Decomposition is therefore very slow, and the accumulation of plant remains leads to the formation of peat. Nearly the whole of the area is now kept free of water by means of pumps so that it may be cultivated; in a few small reserves only may the natural vegetation be studied, and even then allowances must be made before the flora of the predrainage era is deduced from the present artificially natural state of affairs.

Wicken, which for long has been National Trust property is the best-known fen, thanks to the researches of Dr H. Godwin and other workers from the Cambridge Botany School. *Phragmites communis*, the common or Norfolk reed, is the dominant species of the reedswamp, and persists for a long time after swamp conditions have been superseded by fen. The first dominant fen species is the fen sedge, *Cladium mariscus*, which forms a dense mat excluding all but a small number of other species. The soil level is built up rather quickly and, when it reaches a point where it is above the average winter water-level, colonization by bushes begins. The alder buckthorn, *Frangula alnus*, is the most abundant species of bush, but there are three others which are fairly plentiful. For some unknown reason alder is scarce in Wicken, though it is a dominant species in the fens which develop around the broads in Norfolk.

Cladium and other fen plants maintain themselves between the bushes for a long time but eventually are forced out as the bushes grow up and the canopy closes. Their place is ultimately taken by species which are able to live in deep shade.

The bush stage is known as carr and is believed to lead on to oak-wood; but this, though highly probable, is not an observation of fact.

The succession over a great part of Wicken Fen has been deflected by man, who cuts *Phragmites* for thatching, *Cladium* for thatching or kindling, and *Molinia coerulea* (purple moorgrass or flying bent) for 'litter', that is bedding for cattle. If the *Cladium* is cut every four years the establishment of bushes is prevented, and the *Cladium* persists beyond a stage at which it would otherwise have disappeared. But its growth is not as vigorous as when it is left undisturbed and its dominance is less complete; *Molinia* becomes an important member of the community, which the fenman refers to as 'mixed sedge'. *Phragmites* also persists and there are other characteristic species.

If the cutting is annual, *Cladium* does not survive, and *Molinia* comes to dominate the vegetation which, in the fenman's parlance, is now known as 'litter'. *Phragmites* is also a member of this community, in which *Carex panicea* (carnation grass) and *Juncus subnodulosus* (obtuse rush) are subdominant.

Wicken Fen was divided into strips each under different ownership, and different people cropped their plots according to different rotas. The fen, as a result, presented a mosaic of plant communities, and Godwin and his co-workers were able to show that the nature of each one was related to the treatment it had received.

Unspoiled fenland is found round the Norfolk Broads, and on a small scale this part of the country today is probably what great areas of fenland were like 400 years ago. The plant succession is, in general, similar to that just described though it tends to be rather more variable. One notable difference, the dominance of alder in the carr, has already been alluded to. Well-developed fenland also surrounds Lough Neagh in Ireland, where the general picture is similar, though there appears to be a rather more clearly defined zonation at different water-levels. But fen vegetation is not confined to regions with a calcareous water. Pearsall has described a typical fen community at the head of Esthwaite Water, one of the Lake District lakes, and it appears that in this case the copious supply of fine silt is the factor responsible.

Finally there is the formation which in Ireland is known as a bog and in Scotland as a moss. Tansley decided that the two are synonymous, but in his usage he inclines towards bog. The present authors, without wishing to offend national susceptibilities, also favour the word bog. They can then refer to the characteristic plant, *Sphagnum*, by its common English name, bogmoss; otherwise they would have to call it mossmoss.

Sphagnum is almost invariably present where bog conditions prevail and is often dominant; but there are a number of other species which occupy an important place in the community according to small variations in the conditions. Familiar names are cottongrass, beak-sedge, bog-rush, crowberry, and bilberry. Less abundant but probably equally familiar species are sundew, butterwort, and bog asphodel.

Bog conditions will be the ultimate phase of any piece of water which is deficient in bases. Any mountain tarn will eventually pass through this stage; a good example in the Lake District is Blindtarn Moss above Grasmere. Fen conditions in the Lake District depend on a good supply of silt and not on a base in solution and therefore, as the fen is built up, the edges will gradually be cut off from a supply of silt, and fen plants will be replaced by bog species. In Esthwaite north fen today there is a thick growth of *Sphagnum* at the edges and it is gradually encroaching on to the zone where, a decade or so ago, sallows and other bushes were common. Any waterlogged depression in base-poor areas (i.e. hard rocks in the north and west, and sandy heath in the south) will tend to develop bog vegetation. Tansley recognizes three types of bog – valley bog, raised bog, and blanket bog. Everything so far considered is valley bog. It depends for its existence on impeded drainage, and it never loses contact with the surface water. Although austere, it is not quite so austere a habitat as the other two, for the most barren soil yields some dissolved salts.

In the fens of East Anglia patches of *Sphagnum* may occasionally be found on top of tufts of sedge or in other elevated situations. Calcium is unfavourable to *Sphagnum* which, in consequence, is found in such regions only where it is out of contact with the ground water. In East Anglia development goes no further than this, and *Sphagnum* is visited as a botanical curiosity. But in similar situations in Ireland development does proceed, and it is probably the greater rainfall and higher humidity which makes the growth of *Sphagnum* possible. Starting from some raised tussock it spreads outward and upwards until eventually it may smother a region of calcareous fenland; many of the bogs of central Ireland probably had such an origin. The product of this process is known as a raised bog, for it grows up into a rounded hump, which may be of considerable dimensions. The characteristic feature is that it is watered only by rain, mist, and dew. It is, therefore, an extremely rigorous habitat and certain species, notably some of the reeds and sedges, which occur in valley bog, are not found on raised bog. There is a constant building up and breaking down accompanied by a succession of plant species and the sequences have been studied in great detail. A neat round hump develops, and the top may become so dry that heather grows upon it. Then, for some reason, the bog becomes unstable, erosion starts at the top, and pools are formed. These pools extend under the influence of the wind, and eventually eat their way to a point where they run down the hill, leaving miniature valleys behind them. They carry down much peaty matter, which spills over and smothers the vegetation at the edge of the bog, thereby widening the base on which it can develop. Then the bare scars left by these pools are colonized by plants, and a phase of regeneration sets in. Building up and erosion alternate in this way, while all the time the bog is growing bigger and bigger.

Raised bog may start on a valley bog if climatic conditions are suitable, and it seems probable that origin from such a site is easier than origin from a calcareous base. The climate of Ireland, Scotland, Wales, and North England is suitable for the formation of raised bogs. In the central limestone plain of Ireland, where they are known as red bogs, good examples may be seen, but in the other countries good types are hard to find as most of them have been drained by man.

In a moderately humid climate raised bogs develop only in damp places such as valley bog or fen; but in the west of Ireland and the west of Scotland the humidity of the air remains at such a high level throughout the year that bog develops almost everywhere where the ground is fairly level. This type of bog, Tansley's third category, is known as blanket bog. It is found also at high altitudes on Dartmoor and the Pennines, where the increased rainfall and lower temperature of the greater height probably compensate for the lower average humidity. Blanket bog has been less affected by man than any of the other types of vegetation considered in this chapter.

The above account of the plants of swamps, marshes, fens, and bogs, sketchy though it is, shows a series of relationships between the plant and its environment. These relationships are apt to be more definite with plants than with animals living around the water. Indeed when it comes to the birds, mammals, reptiles, and amphibians, with which the rest of this chapter is concerned, it is often difficult to decide which kinds qualify for inclusion. Among the birds, for example, one would not usually think of the swallows as dependent on water; but consider the following observations which show how two broods of swallows could fairly be considered in the productivity balance sheet of a tarn in the Lake District.

A floating cage covering one-third of a square metre of the surface of Three Dubs Tarn, a typical artificial fish-pond, was moored out in the middle, and, in the season of 1947, caught 2,996 Chironomids and other insects which had emerged from larvae dwelling in the bottom mud. In the shallower water, where vegetation was thicker, there were fewer Chironomids but more larger creatures such as dragon-flies and caddisflies; the total catch was 1,993 individuals. The mean of these two figures may fairly be taken as a basis from which to calculate the total number of insects emerging from the tarn during the season of 1947. The surface area is 16,188 square metres and so the final answer is 121,167,180 or 121 million to approximate to the nearest million. A pair of swallows nest in the boathouse of Three Dubs Tarn every year, and in 1947 they raised two broods. The parent birds may be seen hawking continually over the tarn and they doubtless feed extensively, perhaps entirely, on insects which have spent all their growing life in the water. In fact, in this instance, swallows are a highly important part of the animal association dependent on the water.

There are other species of insect-eating birds which may, in particular circumstances, be closely connected with the productivity and the food-chains of fresh water, but in this chapter we have included only those which feed extensively on the products of fresh water and are to be found on water for the whole year or the greater part of it. Visitors on migration do not qualify, and the ducks and geese are treated sketchily.

The other vertebrate animals (other than fish) associated closely with fresh water comprise three mammals and eight amphibians. There is no reptile in Britain which can be included as aquatic or semi-aquatic, although the grass-snake, *Natrix natrix*, is often seen near water or even in water swimming with considerable dash. This is a modest list, which has remained modest in spite of the many doubtless well-meaning but usually ill-advised attempts to enlarge it. At the present moment there is a fourth aquatic mammal, the coypu or nutria, at large in this country, and the memory of a fifth, the musk-rat, is far from dim. On the debit side is the beaver, *Castor fiber*, which was exterminated from Britain some time about the twelfth century. It was once common and its skeleton is found quite frequently in fenland peat and other deposits. But it was a large animal, like many other semi-aquatic mammals it has thick soft fur which was highly prized by man, and it dwells in colonies and so was more conspicuous than solitary species. These three factors doubtless contributed to its extinction from Britain as man gradually cleared and drained the lower courses of the rivers where it lived.

The three native aquatic mammals are the otter (*Lutra lutra*); the water-vole (*Arvicola amphibius*) and the water-shrew (*Neomys fodiens*). The otter, the fox (*Vulpes vulpes*) and the badger (*Meles meles*) are the only moderate-sized Carnivora which have managed to avoid complete or almost complete extinction in Britain. All three sleep underground and hunt by night, and this, coupled with a keen wariness, is probably the reason for their success. But the fox and the otter are also pursued for sport and by virtue of this enjoy some immunity from other methods of destruction. The otter is widely distributed in Britain. It haunts the larger lakes and lower reaches of rivers, lying secure by day in a burrow, technically a holt, which is usually in the bank of a river, but may be at some distance from the water. In north Scotland and in Cornwall otters dwell in caves along the shore and hunt in the sea.

The main diet is fish, which the otter pursues with agility and skill, but its tastes, like those of most carnivores, are catholic, and it will feed on other water animals such as crayfish and frogs, and is stated to be a great enemy of moorhen chicks. It may range inland and go after rabbits and sometimes it raids poultry yards. Many fishermen regard it as an arch-enemy but Vesey-Fitzgerald (1946) states that, since it feeds mainly on eels and coarse fish which compete with game fish for food, in a salmon river it does more good than harm. But in a pond or stream devoted exclusively to trout it can do great damage, and, if in a frolicsome mood or giving its young some tuition, it may destroy a large proportion of the population in one night, killing far more than it can eat. But no lover of beauty could wish the otter to follow the beaver to extinction from the British Isles.

A dog otter, which is larger than its mate, may attain a weight of nearly two stone, though this is exceptional, and there is considerable variation in weight. There is only one litter in the year, and the cubs, which may number five though fewer is usual, are born any time from spring to high summer.

The water-vole is a rodent. It is sometimes erroneously referred to as the water-rat but, like the other two British voles, it has a shorter and blunter muzzle and a shorter tail than the rats. Further, its ears are almost concealed in the fur of the body and do not stand out like those of rats. It is common and widely distributed in England, Scotland, and Wales, though unknown in Ireland, and is to be found in the slower reaches of streams and rivers; it inhabits places where there is plenty of vegetation and soft earth banks into which it may burrow. Its tunnelling habits make it unpopular in artificial ponds with earth dams, and in places where canals and drainage ditches are banked up above the level of the land which they traverse.

The water-vole is a vegetarian, and may frequently be seen sitting up on a patch of vegetation in midstream nibbling a young shoot which it holds in its fore paws. If scared, it drops into the water and disappears, for it is a good diver and swimmer. It is not encouraged in places where water-cress is grown, and it may also damage osier beds, but otherwise it is harmless, though it has been known to wander inland and attack root crops when hard pressed. Unfortunately it is sometimes mistaken for the brown rat and persecuted in consequence. Brown rats may dwell in a stream bank in burrows very like those of a water-vole colony but their feeding habits are quite different.

The water-vole is the largest of the British voles. Breeding starts in spring and several litters are produced before reproduction stops in the autumn. The young number six or seven in the first litter but fewer in later ones.

The shrews may be distinguished at once from the voles, rats, and mice, by the long tapering snout. If the mouth be opened, a more fundamental difference is seen. The voles and their allies have large characteristic chisel-like teeth at the front of the mouth, and there is a gap between them and the grinding teeth farther back. The shrews have a mouthful of teeth rather like our own, though there are more of them. The incisors are quite small and the canines or eye-teeth are large; then, without any gap, follow the premolars and molars, each with several sharp cusps. The shrews belong to the order Insectivora, and the others to the order Rodentia.

The water-shrew is the largest of the three British species and it has stiff hairs fringing the tail and feet. The hairs on the body give it a characteristic broad rectangular appearance under water. The back is black or blue-black, sometimes with a brownish tinge, and the belly ranges in shade from dirty white to pure white; some specimens have white ear-tufts. It is not as common as the water-vole and it is absent from Ireland. Like the vole it lives in burrows in the bank and sallies forth for food into the water, in which medium it is a capable performer. But, unlike the vole, it is a predator and seeks out caddis larvae, snails, and other invertebrate animals on the bottom. It may occasionally attack the spawn and fry of salmon and trout; there are probably few predators of which this cannot be said. Weasels are stated to hunt it assiduously and it frequently falls a victim to the pike.

The female water-shrew constructs a nest at the end of a long burrow and there, in May, she brings forth six to eight young. There may be two more families in the season.

The musk-rat *Ondatra zibethica* is a rodent rather larger than the water-vole, and its natural homeland is Canada. It has inhabited parts of eastern Europe for many years, having been introduced originally on account of its fur, and has recently become established as a wild animal in several parts of western Europe. It has proved a doubtful asset in some places at least, as it creates an extensive series of burrows and, if the site is a canal bank or the earth dam of a pond, its tunnels may undermine the structure sufficiently to cause collapse.

The proposal to introduce the animal into Britain after the First World War was vigorously opposed in certain quarters but efforts to induce the Government to forbid its importation were not successful, and musk-rat farms were started all over the British Isles. Inevitably some specimens escaped and, by 1933, the creature was well established in the wild state at a number of places. The thickest populations were in Shropshire and Perthshire. The Ministry of Agriculture and Fisheries instituted a vigorous campaign against it, and sought the collaboration of the Bureau of Animal Population of Oxford in order to obtain the necessary knowledge about its biology and natural history. The trapping was successful and within a few years the musk-rat was believed to be extinct in Britain.

As already stated, the musk-rat is a great burrower. Newcomers make fresh holes of their own and will not enlarge any pre-existing water-vole burrow, nor, usually, utilize those made by other musk-rats. The terminus of one burrow was found to be thirty feet from the river's edge. The entrances are frequently several feet beneath the water surface, but, where the water level is liable to fluctuate, there may be a series of adits and exits at different levels. In marshes and places where all the land may be submerged, the musk-rat builds 'houses' of vegetation. Occasionally the young are born in these but the more usual birthplace is underground. The food is almost entirely vegetable, and it seems that the animal has a habit of cutting a food-plant into suitable lengths and then bearing these off to eat in the security of a burrow. Frequently some of the lengths get left behind and it is these floating on the water surface which often betray the presence of the musk-rat. Some animal matter is taken. As any vegetarian knows, it is difficult to feed on plants without taking in some of the animals which are attached to them, and the lower animals which subsist on a vegetable diet must consume much animal matter incidentally. But traces of swan-mussels have been found in the stomachs of musk-rats, and so large an organism, and a mud-dweller at that, must have been taken deliberately.

The first litter makes its appearance in April and is followed by a second and sometimes by a third. The average number of young is seven. The young do not attain sexual maturity in the year of birth, and the musk-rat does not reproduce as rapidly as was once feared. A tendency to wander becomes manifest towards the end of the year, and, in the following spring, when the breeding season starts, there is a second period of restlessness, though this one, unlike the other, seems to affect only the male (Warwick, 1940).

The coypu, *Myocastor coypus*, or nutria, to give it its commercial name, is also a rodent, though of much greater size than any so far mentioned; a stone is no more than an average weight for a full-grown specimen. Like the musk-rat it is a foreigner (though it comes from South and not North America), and it has fur which is valued highly, so that the idea of keeping it in farms in England seemed profitable. The first specimens came in by air in 1929, and within a few years there were some fifty farms in the country, most in the south and south-east. All of them, however, appear to have closed when the war started in 1939. It is scarcely necessary to state that specimens escaped. In Norfolk the big reedswamps provided a good habitat, and the animals became well established.

The coypu feeds on aquatic vegetation and is particularly fond of reeds. As it also destroys these plants by trampling them down, it has reduced the reedy areas quite appreciably in some places where it is numerous. It uses reeds to make a large nest, rather like a swan's nest to look at, and in it the young are born, usually five at a time. But they do not remain long in this exposed situation and within twenty-four hours they have taken to the water. The mother has a peculiar feature in that the teats are situated high on her flanks so that the young can feed while she is swimming. After a short period the young are eating vegetation like their parents. They may have become parents themselves in eight months, but usually the period is nearer a year. Gestation lasts about five months, and breeding apparently takes place all the year round.

The coypu does burrow but its retreats are seldom more than a few feet long. They are often open at both ends, and the mouth is usually at water level. The young do not dig much and frequently seek refuge in the burrows of some other animal. The modern moralist who deplores idleness must find pleasing the not uncommon sequel, for the burrow may prove to belong to a rat, who extends no hospitality to the refugee but treats him as an intruder and as a welcome addition to the larder. The burrowing activities are not, therefore, very serious though they might become so if the population were very dense. The coypu first earned a reputation as a pest in the winter of 1942–43 when it started raiding cabbage patches. Later, damage to sugar-beet plantations was recorded, and the Norfolk War Agricultural Executive Committee instituted a trapping campaign. They too sought the collaboration of the Bureau of Animal Population in order that scientific observations might be made.

After the war the coypu steadily built up its numbers and extended its range, advancing beyond the confines of Norfolk and Suffolk to the neighbouring counties of Holland, Island of Ely, Cambridge, Huntingdon, Bedford, Hertford, and Essex. Its reputation as a pest grew simultaneously, for, as it moved beyond the region of fen and marsh, it found more and more of its food in the fields, particularly those in which sugar beet was growing. Moreover when it is present in large numbers, its burrowing causes damage. In 1960 the grant for clearing areas of rabbits was extended to include coypus too, and most of the outlying colonies were eradicated. In Norfolk and Suffolk, however, further steps were deemed desirable and in 1962 a trapping campaign was inaugurated. It was held to be impracticable to exterminate the animal from the thick fens around the Broads, but it was hoped to reduce the numbers there and to confine it to that area. The trappers were aided by the exceptional length and intensity of the cold spell during the winter and the aim was achieved (Norris, 1967).

The British Amphibia include three newts, three frogs, and two toads; two of the frogs are introduced aliens. All three species of newt are fairly widespread, though the great warty newt, or great-crested newt (*Triturus cristatus* or *palustris*) is most abundant in the south of England, and the palmate newt (*T. helveticus*) is most abundant in higher latitudes and also at higher altitudes. In many parts of the country the common newt, or smooth newt (*T. vulgaris*), is not the commonest species of the three.

During most of the year newts live under stones, in walls, and in other damp places, and they lie hidden all day so that they are rarely seen, except by those deliberately seeking them. They come out at night to feed on various small invertebrate animals. In March or April they travel to some pond, and there, for a month or so, they live an aquatic existence quite different from their previous terrestrial way of life. The male develops a crest down his back and a characteristic and striking pigmentation so that he becomes an ornate and handsome animal. The female is occupied with the production of eggs and each one is laid singly and attached to some object, often the leaf of a plant, which is rolled round the egg and glued to form a protective case. By June the male has begun to lose his crest and his gay colouring, and gradually most of the adults disappear from fresh water. The young emerge from the egg as tadpoles and lead an aquatic life for four or five months. Sometimes winter overtakes them while they are still immature, and then they do not become adult until the following season. The fully-formed newt leaves the water and does not return to it until three or four years later.

No more harmless animal than the newt can be imagined but it is still regarded with fear and horror by many people. The Reverend J. G. Wood, writing a century ago, has some almost unbelievable tales about local superstitions. On one occasion he captured some specimens and transferred them to a cattle trough as a convenient reservoir whence they might be taken at his convenience for detailed observation in the aquarium in his study. But the farmer, when he discovered it, came greatly agitated to the parson and demanded the immediate destruction of the newts which, he asserted, had caused the death of a calf. In vain did Wood point out that the calf in question had never been near the trough, that the newt was a harmless creature possessing no organs of offence, and that, since every horsepond on the farm contained newts, not one beast should be alive if the farmer's reasoning were right.

The common frog, *Rana temporaria*, is a wide-spread and common animal to be found almost anywhere where suitable conditions obtain. It belongs to an order which has only partially mastered the art of existence on dry land and it does not possess an impermeable skin. It must, therefore, always dwell in a damp atmosphere, or it runs the risk of drying up. It is a predaceous animal feeding on a variety of invertebrates, slugs and worms being favourite items in the diet. The capture of these needs no special technique but the frog can also capture flies and other insects which could escape from such a slow-moving animal did it not possess a special device for securing them. The free end of the tongue points down the throat and not towards the outer world like the human organ. It is bilobed at the tip and covered with a sticky saliva. It is extensible and can be projected with great speed to seize some agile creature which is resting at what appears to be a safe distance.

Frogs spend the winter in a torpor either in some hiding-place on land or in the mud at the bottom of some pond. They are one of the first animals to become active in the spring, and mating and egg-laying take place soon after the end of the hibernation period. At this point earlier writers on natural history were wont to insert a lyrical passage about the frog as a harbinger of spring and then pass on. Today, thanks to the researches of Savage (1961), much more is known about the factors which influence the behaviour of frogs when they first emerge from hibernation. The exact time of emergence is difficult to establish but it is probably several weeks before spawning. During that period frogs travel to the ponds where they spawn. The number of frogs arriving at some ponds is greater than at others, and frequently frogs will leave a hibernating-pond which looks to the human eye a perfectly satisfactory place for egg-laying and travel some distance to a spawning-pond. Arriving there, they seek out a few restricted areas where the eggs will be laid. In some years this takes place almost at once, in others there is a pre-spawning period during which, having found the oviposition places, the frogs hang about and, although the males may sometimes mount the females, there is no croaking and no spawning. Then one night the males become vociferous and the females start to lay eggs, and the sudden onset suggests that there is some trigger stimulus which is not apparent to the human observer. The date is not the same each year but in some ponds spawning is rather regularly earlier or later than in neighbouring ponds.

Dr and Mrs Savage made detailed observations one year in one particular pond; they chose a spawning-pond which was not as popular as many of the others, so that the number of animals which they had to handle was not too great. The spawning-season lasted twelve days, from 22nd March to 2nd April. Each night the observers captured for quick observation all the frogs in the pond. Any newcomers were marked with a label tied round their middles before return to the water, and the label number of any which had been marked on a previous night was noted. At the beginning there were many males present, and they remained in the pond for several days at least, some being successful in securing more than one mate. As the days wore on, some of the males departed and a few late-comers arrived. Every night a few females came to the pond; they were seized by a mate, they laid their eggs, and then they departed; no female was ever found twice. It was presumed from this that the behaviour in an aquarium, where a male may remain clinging to a female for several days on end, is abnormal; it was demonstrated that, if such a pair be transferred to water taken from a spawning-pond, oviposition takes place within a short space of time.

The female frog has horny granules on the skin and the human observers found that they could easily distinguish a female frog in the dark by touch. A female, who, having laid her eggs, is bound for the bank, will croak at any male who tries to seize her, whereas, before the eggs have been shed, the female never uses her voice. Her thin condition also serves, it is thought, to indicate to the male that he has not found a suitable mate.

These are the observed facts and they present several interesting problems. First there is the question of what determines the spawning time, and also the spawning order of a series of ponds. Savage obtained the co-operation of observers all over the country and was able to gather a great deal of data. It became apparent quite soon that the climatic conditions on the night when spawn is first laid are unimportant. It took place on warm nights, on cold nights, even including nights with frost, on wet nights, and on dry nights. The data gathered, Savage settled down to a long and ingenious analysis. He plotted on a map of the British Isles what he calls isophenes, that is lines joining all the areas where spawning date is the same, spawning date being in number of days after January 1st. Spawning in January, that is within the isophene 30 days, occurs only in Devon and a corresponding area in Wales. Most of Britain from Sussex and Hants to Caithness and

Sutherland, lies between the isophenes 60 and 70. In areas to the west spawning is generally earlier and to the east later, but spawning is also late in the industrial areas of the midlands. The latest spawning of all, between day 90 and day 100, is recorded on Cross Fell in the Pennines. The next step was to plot isophenes against rainfall along the horizontal and temperature along the vertical axis for the month of egg-laying (M_0), the month before that (M_1) and the month before that (M_2).

If M_2 is cold and dry spawning will be late. Below 6° C., increased rainfall advances the date of spawning but not by much. Above 6° C. the higher the temperature the earlier the spawning and the date is not much affected by rainfall. If M_1 is very dry spawning will be late, whatever the temperature. More rain means earlier spawning, but temperature exerts an influence as well. A rise in temperature up to 0° C. is followed by earlier spawning and thereafter up to 3° C. by later spawning. Increasing warmth brings the spawning time forward once more up to between 7° and 8° C. after which it retards it. In a very dry M^0 spawning is earlier than in one moderately dry with a rainfall of about 80 mm. Otherwise the pattern resembles that in M_1. There is also evidence that light has an influence on spawning date.

It is extremely unlikely that temperature, rainfall, and sunshine have much influence on frogs hibernating in the uniform conditions in a pond. The second and third, however, are of much significance to algae, rainfall washing in nutrients and sunlight providing energy. It is probable that the wax and wane of different species of algae varies from place to place in one year, and from year to year in one place under the indirect influence of these two climatic variables and possibly temperature also. Savage postulates that algae are the links between the weather and the frogs. That certain species of algae have a characteristic smell is well known to waterworks engineers, and perhaps frogs choose a spawning-pond by its smell. There is probably no direct connection between any one species of alga and the nutrition of the tadpole; but possibly a certain alga is associated in the frog's nervous system with suitable conditions for egg-laying and the subsequent development of the young. Furthermore, Savage puts out the suggestion that the exact spawning-locality in a pond is chosen on account of some algal feature, and that the stimulus which starts spawning is some vegetational change. All this correlation with plants is hypothesis at present; but it is in accord with the facts, and will provide a starting-point for investigation when a team of the right combination tackles the problem.

Frogs' eggs hatch about a week after they have been laid, the exact length of the incubation period depending on the temperature. At first the tadpoles remain congregated together but eventually they disperse to seek their food. Here we encounter a gap in knowledge, not only about tadpoles but about many other animals as well. There is plenty of information about what they eat, much less about what they can make use of, and no doubt that the two do not coincide. Savage (1961) records that, though the gut of a tadpole is never empty, growth proceeds at a variable rate and sometimes stops entirely. Passage of food through the gut is rapid and much is often obviously undigested. There appears to be no means of digesting cellulose, yet tadpoles feed extensively on algae and other vegetable matter. Moreover they grew well for a time when fed on cultures of one species of alga. Savage suggests that the vigorous peristaltic movements of the gut rupture the cell walls of the algae and expose the contents to the digestive enzymes. Any herbivore consumes some animal matter and Savage noted that some tadpoles were feeding mainly on water-fleas. In the pond where this was observed growth was good, and it seems likely that animal tissue is important in the diet of all tadpoles. The tadpoles are preyed on by almost every predator, and one may imagine their soft bodies make tasty and nutritious eating. One of the authors has seen eight *Hydroporus* beetles set upon a tadpole several times larger even than all the attackers put together, and eventually by combined efforts, like a pack of wolves, overpower and devour it.

After an aquatic life of nearly three months the tadpole undergoes a remarkable metamorphosis, changing from a swimming aquatic herbivore to a hopping terrestrial carnivore; but the details are too well known to require repetition.

The edible frog, *Rana esculenta*, is not indigenous to Britain but has been introduced many times and is now established in some parts of the fens, and around London, though it has never spread very far from the original point of release, and has died out in several places where it was once known.

The marsh frog, *Rana ridibunda*, resembles the edible frog but is larger. Mr E. P. Smith, who thought at the time that it was the edible frog, was the first person to set it free in this country, and he has given an account of its early years at large in the *Journal of Animal Ecology* for 1939. In 1935 some specimens which he had obtained the previous winter bred in an ornamental pond in his garden in East Kent, though the breeding was not very successful, since most of the tadpoles fell a prey to fish. In June two individuals migrated to a mere about half a mile away and in October the rest followed them. Wandering in these two months, but particularly in October, appears to be characteristic of the species.

In 1936 the frogs bred in the mere and produced a great many young, their success being probably due to the fact that all the predator fish had perished in a drought two years previously. There were a few records of specimens at various points round about, and there were a few complaints about the noise of the frogs croaking at night.

In 1937 a peak was reached. The frogs spread over an area which was roughly a square having sides about twenty-eight miles long. Most of the area was old marshland intersected by dykes and drains, but the frogs appeared in one lake on the uplands fourteen miles distant from the mere and unconnected with the marsh area. The noise set up by the frogs at night was so great that nobody living in the vicinity could sleep. It is said that one of the tyrannies which drove the French peasants to rise up and start the French Revolution was that they had to spend the night flogging the water to keep the frogs quiet whenever aristocracy saw fit to sleep in the country. The frogs of East Kent also raised strong feelings, and a parish meeting of protest sent complaints to the Ministry of Health and to the local M.P. But both the Ministry and Parliament behaved much as did the House of Peers when Wellington thrashed Bonaparte. The daily papers, needless to say, made the most of the phenomenon. The result was unexpected. A brisk demand for frogs came from other parts of the country. The reason is not stated, except that one place ordered about 200 to cope with a plague of mosquitoes; a surprising purchase, for it is unlikely that frogs of any kind have any effect on populations of mosquitoes.

In 1938 it appeared that the zenith was passed. The frogs were still numerous but something had checked the wholesale increase of the year before; it scarely seemed likely that the activities of the salesmen were on a large enough scale to be the cause. But most unaccountable of all was the fact that the frogs were silent. The whole story remains a mystery. Twelve specimens were introduced originally; after three years it appeared as if they might flourish like the rabbit in Australia, and their vocal efforts constituted a major nuisance. Then quite suddenly their numbers began to dwindle and their song ceased. That was the situation as it appeared to Smith at the time. Actually the check was probably not as great as he supposed, for, in 1949 in a personal communication which he was kind enough to make, he writes that the frogs are more abundant than ever, though not as noisy as formerly.

The common toad, *Bufo bufo*, is encouraged by most people today because in a garden it devours slugs and other animals in disfavour with the cultivator. Like the frog it must find some damp retreat into which to withdraw when the sun is up, for it cannot control the loss of moisture from its surface, though it does not lose water as rapidly as the frog. But the idea that it is a beneficial creature is of recent growth, and the wildest beliefs about its venomous nature are even now hardly dead in the more backward parts of the country. The Reverend J. G. Wood describes a scene which one cannot picture without a smile. He was walking with some French friends in France. One of them spotted a toad beside the pathway and, recoiling in horror, he raised his stick to strike the animal dead. But before the blow could descend the Englishman had rushed in and seized the uplifted arm. Then Wood bent down to pick up the toad, and now it was his turn to be seized and forcibly restrained; nothing could convince the Frenchmen that painful death would not follow the touching of a toad. Finally it was agreed to sprinkle some tobacco on the toad's back, for the Frenchmen were convinced that this would kill the animal, and Wood knew quite well that it would not. And so the party proceeded, a compromise having been reached and international unity restored.

The toad usually spawns a fortnight or more later than the frog, frequently choosing large bodies of water. Pairs are found in Windermere and certain of the perch traps, about which there is more in Chapter 12, catch numbers of them every year, even at a depth of twenty feet. The process of egg-laying takes much longer than in the frog, and during the course of it the eggs, which are attached together in a long string, become wound round plants and other obstructions in the water. When all the eggs have been laid, the female repeats certain movements which she made at the beginning of spawning but no eggs come forth, and at this the male releases her. Any male which may seize her subsequently is dismissed in the same way.

The natterjack toad, *Bufo calamita*, which is smaller and has a light stripe down its back, is restricted to a few parts of the country, though where it does occur it may be numerous. The neighbourhood of ponds in the slacks between sand-dunes is a favourite haunt.

The water-beasts, being so few in number, have been considered one by one roughly according to the order in which they come in the zoological scheme of classification. The water-birds are more numerous, and many of them can be related more definitely to the environment. Therefore it is preferable to take them according to their habitats, starting, as before, with Ennerdale Lake and passing through a series to rich lowland localities.

The most primitive, Ennerdale, is an inhospitable lake for birds and, though many species may be seen upon it when migration, hard weather, or rough seas bring them from their usual feeding-grounds and resting-places, there are but two which make a regular living from it. One is the dipper, a species more typical of streams, and the other is the cormorant, of which one or two can usually be seen sitting in security on the small pile of stones which constitutes an island in the middle of the lake. This bird does not breed inland in the Lake District, though there are records of it doing so elsewhere, but repairs to the sea coast where it nests in colonies on cliffs. A few birds haunt the lakes all through the year, and numbers inland increase during the winter. It is almost exclusively a fish-eater, and in Ennerdale presumably preys on trout and char, because there are no other fish for it to eat. It is one of the few birds capable of catching fish in the deep and coverless conditions of Ennerdale, for it can swim under water with the aid of its legs and wings with remarkable agility. A trout weighing 2 lb. has been taken from the maw of a cormorant and it is regarded as a most undesirable bird by most fishermen.

On Windermere, however, the cormorant is probably beneficial, for it is one of the factors operating to keep the excessive population of perch in check, its diet consisting largely of this fish. Mr P. H. T. Hartley has studied its behaviour and has collected together the results of other ornithologists. The cormorant, it seems, feeds three times a day at very much the hours when man is wont to take his breakfast, lunch, and tea. When not feeding, the birds sit on small rocky islands out in the lake, and so it is easy, if tedious, to make out how often and when they feed. The average weight of fish per meal is probably about half a pound and, as there are about twenty birds hunting the lake, upwards of five tons of fish are consumed by them during the course of a year. To the local inhabitants the cormorant is a fish-eater and therefore a villain, and they shoot him when they can; which is not very often, for he keeps the lake swept with a wary eye from his post of vantage, and seems to know more about the range and accuracy of shotguns and rifles than many of the people behind them.

Windermere is not a favourable place for a permanent bird population for two reasons. First there is no extensive development of reed-beds, and secondly throughout the summer there are a great many humans in pleasure boats. The most conspicuous bird is the mute swan, though one may doubt from its behaviour whether it is really to be classed as a wild species. According to the text-books the swan feeds almost entirely on vegetable matter, but on Windermere for several months of the year its diet includes a wide range of animal, vegetable, and mineral matter, for the swans feed readily on whatever they can beg from those enjoying a picnic beside the water or dwelling in a houseboat or cruiser. It is also alleged that swans consume a lot of the char spawn, which is deposited in shallow water in winter-time.

The nest is usually in a reed-bed, though sometimes one of the islets is chosen. No secrecy surrounds the nesting of the swan; well-trampled reeds, a vast mound of dead reed-stems making up the nest, and the colour of the bird itself advertise what is afoot for all to see. Fortunately for the swans most of the human population believe that one blow from a swan's wing will break a man's leg, and that any intruder is likely to be flung back into the water thus disabled and left to drown. An untimely rise in the lake level is more likely than the human robber to bring the reproductive efforts to naught.

A few ducks and other birds may breed on Windermere in the reed-beds or in the woods near the water's edge, but their nest-making is unobtrusive, and, by the time the lake is well thronged with holiday-makers, the parents and the young have departed to some less frequented stretch of water. The ornithologist will follow them perhaps to Esthwaite or Blelham Tarn which, besides being less haunted by man, have also more extensive reed-beds.

Several pairs of great crested grebes nest on Esthwaite, and the species is also to be seen on Blelham. In April it selects some suitable site well hidden in the reeds and constructs a nest of reed-straw, partly floating and partly supported by the reeds. Three to five eggs are laid in May, and there is usually but a single brood in the season. The grebes are expert swimmers and divers, and capture small fish and also invertebrate animals.

The little grebe or dabchick is also a characteristic species of reed-fringed lakes though it tends to move to smaller bodies of water to breed. The nest is a heap of vegetation, and, like its larger relative, the bird has the habit of covering the eggs whenever it leaves them. The young take to the water at once and soon become expert divers, disappearing below the surface not only in search of food but also when danger threatens. In the earliest days a parent may seek safety below the surface with a young one clinging to its back. The point of reappearance may be a considerable distance from the point of disappearance; as much as forty yards has been recorded. The little grebe feeds on small fish such as sticklebacks and on invertebrates; it may itself fall a victim to a pike.

A third species associated with this type of lake is the coot, a bird which is gregarious and often to be seen in large numbers, particularly in winter when it tends to gather in the lowlands on the largest stretches of water. It nests in reeds or on land at the water's edge, and lays six to nine eggs, sometimes as early as March. It is also a diver but, unlike the two preceding birds, it is mainly a vegetarian; but any animal feeding on pond-weeds cannot avoid consuming large quantities of animals as well.

On a calcareous eutrophic lake, as compared with an oligotrophic one, there is no sharp difference in the bird fauna as there is in the invertebrate fauna, but a lowland lake with a basin so shallow that weeds can grow everywhere and with thick reed-beds all round the edge will support a much greater population of birds than a deep oligotrophic lake in a mountain district. The optimum habitat was probably the open mere, which was a feature of the fenlands before they were drained. Now these have gone and the effect has been great, for aquatic birds are adapted rather than adaptable, and, unlike many woodland species, they have not been able to colonize new types of country when their typical haunts have been destroyed.

But it is on the truly fen species that man's activities have fallen hardest. Cranes, spoonbills, and Savi's warblers breed no more in the British fens, and several other species are, or have been, on the verge of extinction. But, during the last forty years, the setting aside of reserves and special measures to protect rare species during the nesting-season have achieved good results. The ruff bred regularly till 1871, but since then there have been only a few records of nests and the status of the species is doubtful. The breeding status of the marsh-harrier is still dangerously weak, though that of Montagu's harrier is better, and improving. The black-tailed godwit, almost extinct in 1829, has possibly re-established itself as a regular breeder, and the bittern and avocet certainly have. The bearded tit concludes the list of species which have probably been saved from extinction.

Rivers and lakes cannot be separated clearly according to their bird fauna, but most naturalists will tend to associate certain species with either the one or the other, and it is convenient to keep them apart as has been done hitherto. But it must be emphasized that we are not seeking to make a distinction between river birds and lake birds.

Mountain streams and rapid reaches of the foothills are haunted by the dipper or water-ouzel, but where the flow is sluggish this species disappears. Accordingly it is familiar in the north and west of England and in Scotland and Wales, but a rarity in south and east England. It is a distinctive bird with its wren-like shape, its white breast, its habits of perching on a stone in midstream and making curtseying movements, and of flying straight up the river close to the water surface uttering a high-pitched whistle as it goes. It ranges almost to the top of the Pennines and nests up to 1,700 feet, but in winter it withdraws to lower altitudes and frequently establishes a beat along some suitable stretch of lake shore. Breeding starts early in the year, in March or the beginning of April. The nest is built, if possible, on some vertical face either on a bank or a slab of rock or beneath a bridge. The bird has been known to make use of holes and there are records of nests located behind waterfalls. The nest is usually a carefully built domed structure like that of a wren, but larger. The eggs number four or five and there are two and occasionally three broods in the year. The food is mainly aquatic invertebrates, which it pursues under water, either swimming by means of its wings, or, more usually, walking along the bottom. It overcomes its natural buoyancy by spreading its wings at the appropriate angle so that the current presses its body downwards. Flying insects are sometimes taken in the air.

Farther down a stream, where the water flows less turbulently, there will be a reach which is the territory of the kingfisher, another bird found usually beside running water and rather exceptionally around ponds and lakes. The nest is at the end of a tunnel in some soft earthy bank and consists mainly of regurgitated fish-bones. There are about seven eggs and they are laid towards the end of April, with sometimes a second brood in the course of the summer. The bird is a fisher which secures its prey by diving. It commonly has a perch commanding a view of some suitable stretch of river, from which it dives obliquely into the water, when a minnow or stickleback or the young of some larger fish comes into view. Sometimes it dives from a hover. Insects and other invertebrates are also eaten.

Another bird of the lowland river is the heron, though it is to be found anywhere where there are suitable hunting-grounds. It is a gregarious nester though many of the colonies contain only three or four nests. They are usually in the tops of tall trees, and the business of nesting starts early in the year, sometimes before March. The heron is a fisher and predator, but its method is unlike that of either of the two preceding species and indeed unlike that of any other bird so far mentioned. It selects some suitable shallow and there it stands motionless, or stalks forward slowly and deliberately, until some creature comes within reach of its long sharp bill. Frogs, water-voles, birds, especially young chicks, crayfish and other invertebrates all fall a prey to the heron, but it is primarily a fish-eater. As such it is persecuted by the human fisherman, sometimes unjustly, for it takes many eels and other fish which compete for food with species more prized by the angler, but sometimes with reason, when it is found haunting some stream or pond where there is nothing but game fish. However, we refer our readers on to p. 240, where the question of fish stocks is discussed.

As the river becomes slower and broader and more fringed with reeds along the margin, the number of bird species increases. Dabchicks, coots, mallard, teal and other ducks, swans, and other birds of the large lake may be encountered.

On stagnant water decreasing size goes together with a decreasing number of species until there is but one left, the moorhen or, more appropriately, the water-hen. A pair may be found on quite small ponds, even on those of less than one acre in extent, if there be some cover provided by reeds or overhanging bushes. The nest is made in reeds or in overhanging vegetation, though sometimes it is in a pollard willow or other tree by the water's edge. Five to ten eggs are laid in April, and there may be three broods in the year. The bird can dive and it feeds on aquatic plants, but it also forages on land and consumes grain, fruit, or even grass. Otters are said to eat many moor-hens.

Mallard and teal and, more rarely, some of the other ducks nest on overgrown ponds, but they are but temporary visitors leaving soon after the young can fly. In the winter they congregate on larger sheets of water, sometimes at great distances from the breeding-place. The mallard appears to be a genuine omnivore among birds. It feeds on aquatic weeds and will also seek food in the cornfields or pick berries from the hedgerows; but it is known to eat frogs and frog-spawn and other animals which could not possibly be taken incidentally with vegetable food.

A big pond which has reached the swamp stage is often chosen by black-headed gulls for a nesting-place. Their gulleries may be found in a variety of different situations, on sand dunes, around a group of small ponds, even in trees, but the favourite place is some pond or lake where there is a floating mat of vegetation. Such a situation offers some protection from the depredations of man, who cannot force a boat through the floating plant mat and dare not walk upon it; and some protection is needed, for the screaming of the great concourse of birds proclaims the whereabouts of their breeding-place for miles. The nests are piles of vegetation, and the eggs usually number three; they are laid in late April or early May and a second brood is sometimes reared. The food is varied but chiefly insects. They consume wireworms as the plough exposes them and they also take craneflies; there are records of their congregating to batten on plagues of bracken-clocks and antler-moth caterpillars, so they are to be numbered with the beneficial species. But the fisherman looks askance at their habit of hawking over the water and snapping up mayflies and other emerging insects.

11. Animal Travels

Most authors use the word migration for almost any travelling from one place to another, either of single animals or of flocks of animals. Here the term is used in the restricted sense proposed by Heape (1931) to include only those travels which involve a return journey. A further item in Heape's definition is that the two places of sojourn should be separated by an area in which the animals are never found except when passing through it or over it on migration. The annual journeys of the swallows and the cuckoos provide one of the most familiar examples of migration. The phenomenon is not necessarily connected with reproduction, as it is in these two instances.

Among animals which live in fresh water two fishes, the eel and the salmon, perform migrations which are among the most remarkable feats achieved by any organism. Aristotle knew that at certain times of the year eels go down to the sea, that they have no obvious sexual organs, and that the small elvers which ascend rivers are the young of the eel. The sex organ of the female eel was first described in 1777, that of the male about a hundred years later, and, apart from this, it is true to say that there was no advance on what was known to Aristotle till nearly the beginning of the present century. It might almost be said that there was retreat rather than advance, for some of the theories propounded in the interim were very far from the truth and did not always take into account such facts as were known. It gradually came to be accepted, however, that eels bred somewhere near the mouths of rivers, either in the estuary or in the sea, a fallacy which was not exposed until the present century.

In 1856 a flat transparent fish living a planktonic existence was found near the Straits of Messina. It was not obviously related to any known fish and the genus *Leptocephalus* was created to receive it. Forty years later two Italian workers showed that this creature was in fact a young eel. *Leptocephalus* is, therefore, invalid as a name, but it is convenient to use it as a description and to refer to the leptocephalus larva or leptocephalus stage in the eel's life-history.

In 1904 a Danish worker, Johannes Schmidt, investigating

the life-histories of various fish of commercial importance in the research ship *Thor*, caught a leptocephalus in deep water off the Faeroe Islands. It was a significant event, for it suggested that the commonly accepted view of a shallow-water breeding-place for eels might be wrong. Schmidt undertook to find out where the eel did breed, and committed himself thereby to a quest which was not crowned with success till 1920, sixteen years later.

In 1905 and 1906 the *Thor* sailed to more southerly collecting-stations in the Atlantic, and many leptocephalus larvae

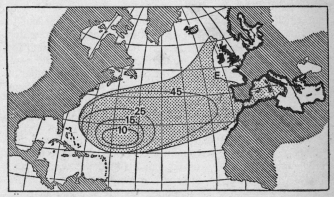

Fig. 23 Life history of the eel. Each line encloses the area beyond which leptocephalus larvae of the size shown do not occur; the figures show length in millimetres. Coasts invaded by elvers are shown black. (After Schmidt, 1922)

were collected. In 1907 there was no voyage and Schmidt turned his attention to the adult eel. He found that its range extended from the north of Norway to the Azores and Egypt (see Fig. 23), and that there was not more than one species, contrary to the views of several previous workers who had recognized a varying number of different species. This lent some support to an idea, which Schmidt was formulating, that all the eels from Europe and the Mediterranean might have a common breeding-place.

Attention was then turned to the Mediterranean and, after many stations had been worked and many specimens examined, Schmidt was able to show that the farther from the

Straits of Gibraltar the bigger were the leptocephalus larvae. This led him to suggest that the Mediterranean eels bred at some place outside the Mediterranean and that the leptocephali returned to that sea from the Atlantic through the Straits of Gibraltar. The Italians had hitherto held almost a monopoly of research on the eel, and they viewed with disfavour the incursion of a foreigner into their field. They deprecated forcefully the suggestion that 'their' eel should breed anywhere but in 'their' sea, and one may surmise that scientific judgment was swamped in the stress of personal and national feelings.

Meanwhile the Norwegian vessel, *Michael Sars*, had collected off the Azores specimens of leptocephali smaller than any which had previously been brought in. It was evident to Schmidt that the solution of his problem demanded a research vessel with a cruising range greater than that of *Thor*, and for this he was constrained to wait. He did not waste time, however, and he set about securing the co-operation of a number of Merchant Service captains, who undertook to make collections for him while at sea on their lawful business.

The result was that Schmidt had at his disposal data from all over the Atlantic and, when in 1913 he set out in a schooner, he was able to go straight to the place where he believed that his quest would end. His findings confirmed his expectations, but he was not quite satisfied that his evidence was as conclusive as it should be, and he was condemned to a further period of waiting, for the schooner was wrecked and then war came to put an end to Atlantic exploration.

At length, in 1920, he was able to set out in the research vessel *Dana*, provided by the brewers of Carlsberg beer. He was able to fill in the details left undisclosed by previous voyages, and he felt justified in committing his findings to print. The English reader will find an account in *Nature* for 1923 and 1924. The most comprehensive treatment of what is known about eels today has been written by Bertin (1956).

Schmidt did not actually see eels laying eggs, and his deductions about the place of oviposition were based on the distribution of larvae of different sizes. Figure 23 shows one of the ways in which he summarized his findings. Each line encloses an area outside which eel larvae below a certain size were only very rarely found, and the figures inserted on the lines show the size in millimetres. Larvae less than 10 mm. long have, except in rare instances, been found only in a restricted area of the Atlantic. This area, Schmidt concluded, is where the eggs

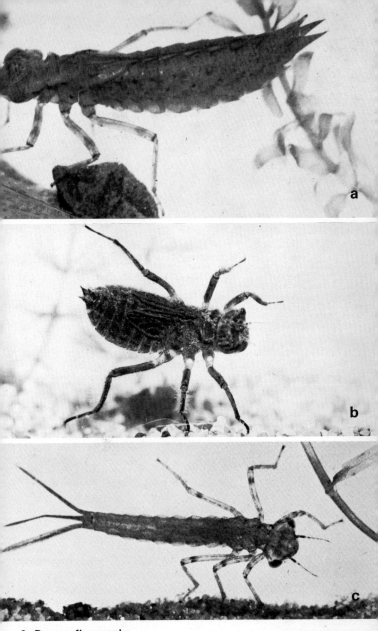

9. Dragon-fly nymphs
a. Aeshnid. *b*. Libellulid. *c*. Agriid

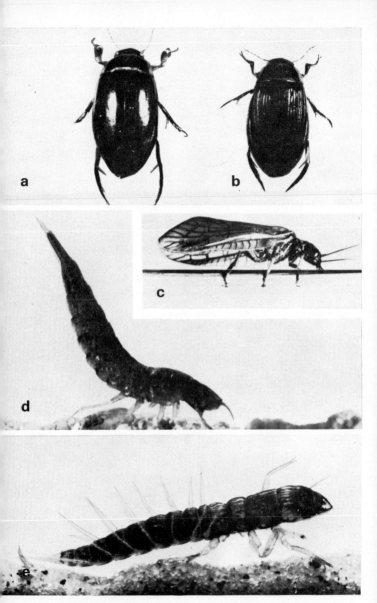

10. Water insects
a. Male (smooth) and *b.* female (furrowed) and *d.* larva of the water beetle, *Dytiscus marginalis*. *c.* Adult and *e.* larva of the alder fly, *Sialis lutaria*

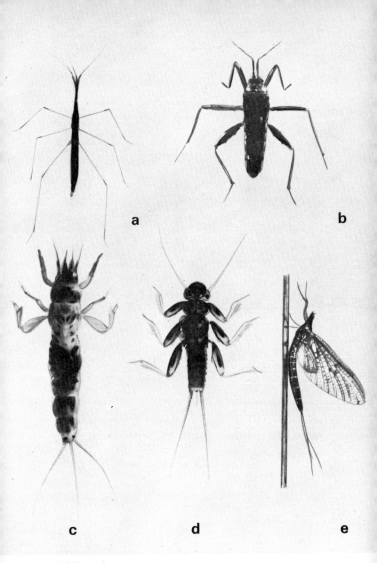

11. Water insects
a. Water measurer, *Hydrometra stagnorum. b.* Water cricket, *Velia currens. c.* Nymph and *e.* adult of *Ephemera danica*, the Mayfly. *d.* A creeper, the nymph of the stonefly, *Perla bipunctata*

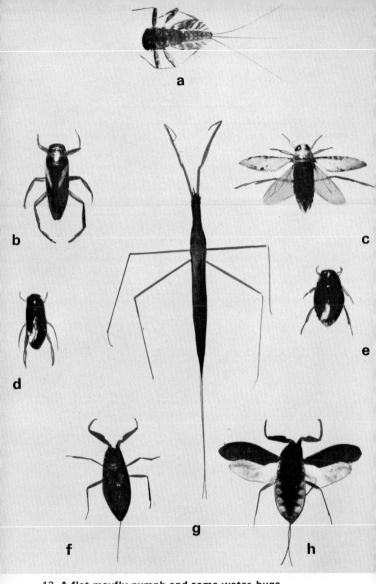

12. A flat mayfly nymph and some water-bugs
a. Ecdyonurus sp. (Ephemeroptera). *b.* to *e.* Water boatmen;
b. Notonecta obliqua (furcata), c. N. Glauca, d. Corixa punctata,
e. Naucoris cimicoides. f. and *h.* Water scorpion, *Nepa cinerea.*
g. Water stick insect, *Ranatra linearis*

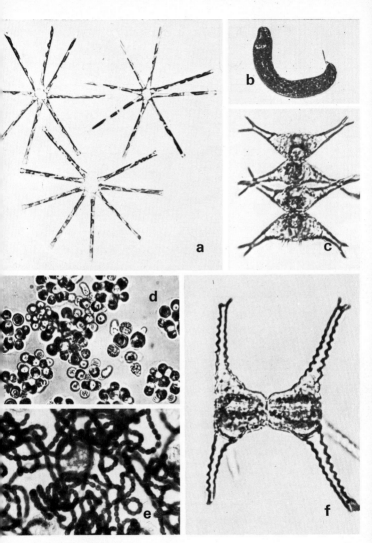

13. Planktonic algae

a. Diatom, *Asterionella formosa*. *b*. Flagellate, *Euglena spirogyrae*.
c. Desmid, *Staurastrum pseudopelagicum*. *d*. Colonial green alga,
Sphaerocystis schroeteri. *e*. Threads of the blue-green alga, *Anabaena
flos-aquae*. *f*. Desmid, *Staurastrum gracile*

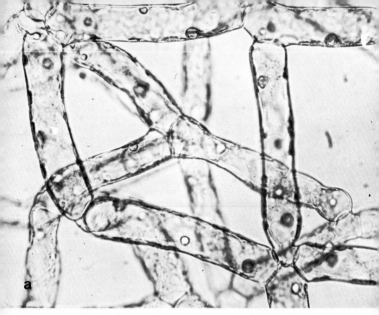

14. Algae

a. Cells of the water-net, *Hydrodictyon reticulatum*. *b*. Threads of the filamentous green alga, *Mougeotia* sp., and zigzag colonies of the diatom, *Tabellaria* sp.

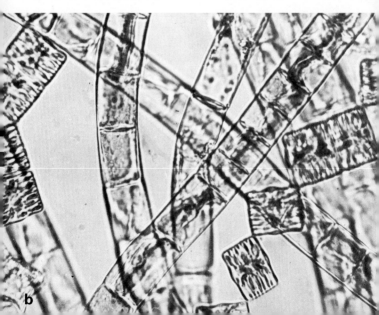

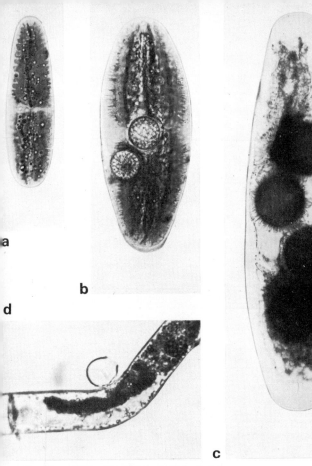

15. Chytridiales, fungal parasites of algae
a. A healthy desmid cell. *b.* A similar cell containing two parasite bodies. *c.* A desmid cell with four parasite bodies now encysted in spiny envelopes among the remains of the cell contents. *d.* A spore-case on the outside of a *Mougeotia* sp. filament

16. Dipper

are laid. The smallest larvae of all are found at a depth of about 400 metres, and so it is probably at about this depth that oviposition takes place. It is near the limit to which any light penetrates, and the particular area is characterized by a rather sharply marked temperature-gradient, and at 400 metres (1,300 feet) is enclosed by the 15° C. isotherm. The ocean floor is 4,500 metres (nearly 15,000 feet) below the surface in this part of the Atlantic.

It is thought that oil globules in the eggs counterbalance them so that they neither rise to the surface nor sink to the bottom.

Egg-laying takes place from March to July, and the newly hatched young swim up to the surface layers of the sea. In the summer in the western Atlantic they abound, but they are gradually swept away to the eastward by the current which flows perpetually towards the shores of Europe. By the following summer they are in mid-Atlantic, and by the summer after that, in which they enter their third year, they are approaching the edge of the European continental shelf. In the winter a change or metamorphosis takes places and the flat transparent leptocephalus turns into an eel-like elver.

Elvers may reach the extreme south-west of Ireland as early as November, though the main invasion does not take place till January. The west coast of England is reached a month later. The coasts exposed to the Atlantic receive the greatest number of elvers, but some drift on, and there is no coast which may not be invaded by at least a few. The remotest corners of the Baltic are reached, though the numbers attaining this destination are small; before the war a German firm operated a station on the Severn to catch elvers, which were then transported to stock north-European rivers.

The elvers which journey on to the farthest destinations celebrate their third birthday on the way, and continue to grow all the time. In the most northerly area of the Baltic, specimens less than 200 mm. long are never seen.

It will be noted from Figure 23 that those larvae which are carried into higher latitudes metamorphose while still far out to sea and leptocephali do not reach the shores of north-European countries, but those that pass into the Mediterranean may travel as far as Italy before all have metamorphosed. The Mediterranean was the place where the original *Leptocephalus* was taken.

Under the influence of fresh water elvers travel actively.

Some turn aside into coastal lagoons and marshes near the estuary, while others run far up the rivers and finally reach the remotest ponds, sometimes after an overland journey made possible by heavy rain. Those which travel far nearly all turn out to be females, and those which find a harbouring place without a long journey nearly all turn out to be males. It is possible that the sex of the eel is not determined genetically, as is that of most animals, but may be influenced, in part at least, by the environment.

The leptocephalus, in its $2\frac{1}{2}$ years of life, feeds on other animals in the plankton, but it makes remarkably poor growth and attains a length of no more than 75 mm., which is about three inches. It fasts for a while during the period of change into the elver stage, some of the tissues are resorbed, and weight is lost; Schmidt estimates that about 1,500 elvers go to 1 lb.

The main feeding and growing period in the life of the eel is spent in fresh water, and during this time it is known as a yellow eel. In due course, after a shorter or longer period according to the temperature, the food supply, and other environmental factors, the eel undergoes further changes. As far as is known, these are related to size, and in Windermere the period is from 9 to 19 years for females and 7 to 12 years for males, which do not grow as large as the females. The changes involve the pigmentation of the skin, the 'yellow' eel becoming a 'silver' eel, and enlargement of the eyes and nostrils.

On some dark and stormy autumn night, with the streams and rivers running high, the silver eels leave the pond or lake where they have dwelt so long and set off on the return journey across the Atlantic. In many other animals the onset of migration is closely correlated either with the development or with the degeneration of the sexual organs, but it seems that the stimulus which sets off the eel must come from some other source, for the sexual organs do not start to develop until it is well on its way.

Not a great deal is known about the eel after it has left fresh water. It starts its sea journey with a body richly stored with food material and, in fact, the flesh of the silver eel comes second only to that of the salmon in food value. Possibly this store is sufficient for the whole of the journey across the Atlantic and no food is taken in the sea, a view supported by the observation that there is considerable degeneration of the alimentary canal. Finally, six months or so after setting out, the eels reach the area in the south-west corner of the Atlantic where the eggs are laid. Nothing is known about the process of oviposition nor of what happens to the parent eels once this has been accomplished. Nobody has ever taken a full-grown eel returning to fresh water and, though the possibility that existence continues in the sea cannot be ruled out, the degeneration of the alimentary tract provides strong support for the belief that the eels die when reproduction has been achieved.

It is an extraordinary life-history and one which confers no benefit upon the species as far as can be made out, and so its significance has been widely discussed. Perhaps the most plausible explanation is that which relates it to Wegener's idea of continental drift. According to this hypothesis, the Atlantic was once a comparatively narrow channel separating the Old World from the New, and the eel, journeying to the sea to breed, had no great distance to go. Gradually America and Europe drifted apart, and presumably the piece of sea with the conditions which the eels chose for oviposition travelled with America and receded from Europe, as the gap between the continents widened. The breeding journey of the eel became longer and longer until today it covers some 3,000 miles. Supporters of this hypothesis point to the prolonged larval life, unknown in any other fish, and to the fact that during this stage growth is extremely slow. What was once a comparatively short and rapid stage has been progressively drawn out as the distance to be covered has steadily waxed greater and greater.

There is much that is remarkable and unexplained in the life-history of the eel, but perhaps nothing more so than how it finds its way across miles and miles of featureless sea to one particular patch of the Atlantic. It drifts westwards with the current. The adults returning must converge on one comparatively small area from all parts of a coastline that extends from inside the Arctic Circle nearly to the Tropic of Cancer. The difficulty of conceiving a mechanism whereby this could be achieved led Tucker (1959) to postulate that all European eels die in the sea and that only American eels return to the spawning ground. Some of their progeny are caught in a current that takes them to America, others in one that carries them to Europe. Early in life these two groups will have been subjected to different temperatures, and this produces the different number of segments and other features which distinguish the two species. Comparable changes can be induced in other fish in this way.

This hypothesis has provoked a violent reaction. For example, Bruun (1963) has demolished a number of arguments which Tucker used to support his contention. Sinha and Jones have launched a more recent attack (1967) and quote some serological work which indicates that the American and European eels are distinct species. These various authors have weakened Tucker's case but not disposed of it, and the controversy is likely to go on until more evidence is forthcoming. If ripe European eels were to be captured on the spawning area, no more would be heard of Tucker's theory. If somebody would produce both species from eggs laid by the American eel, Tucker's case would be greatly strengthened though not proved beyond all doubt.

No less remarkable and even less known is the life-history of the salmon, which is the exact antithesis of the eel, since its main feeding and growing period is spent in the sea and its youth in fresh water. No mystery shrouds the whereabouts of salmon eggs; they are to be found in rivers and large streams where the bottom is of coarse gravel or moderately small stones. They take a long time to hatch, the actual rate of development depending on the water temperature, and they may lie for several months. The young salmon, which are known as parr, are usually active by April or May and feed voraciously on whatever animals they are able to obtain. Growth depends on various factors, notably the supply of food and also the temperature. K. R. Allen believed that the parr are inactive in winter and do not grow at all, the critical temperature being, apparently, about 7° C. Salmon lay their eggs in the colder parts of rivers, where, moreover many of the insects on which the young feed grow during the winter and tide over the summer as an egg. It seems, *a priori*, unlikely that any animal dwelling in these conditions would become inactive at a temperature as high as 7° C., and Jones (1959) has frequently observed parr feeding at 4° or 5° C. and sometimes when the temperature was as low as 1° C. Available to Jones was an underground chamber beside a tributary of the River Dee and separated from it by sheets of plate-glass, and he was able to watch salmon in a way that had not been done before.

As with the eel so with the salmon, size appears to be the chief factor determining the age at which migration to the sea takes place; and there is a further resemblance in that both fish undergo a change in colour before leaving fresh water. The parr loses its characteristic thumb-like black marks down each side and its red spots, and becomes a silver-coloured smolt.

Under exceptionally favourable conditions the change and the departure for the sea take place after one year in fresh water, but smolts are usually two years old and often three or even more. The smolts move downstream, feeding as they go, and eventually in May or early June they reach the headwaters of the estuary, where they congregate. Then one day they all disappear into the sea. As will be described presently, it was not until 1965 that more than an occasional salmon was caught by commercial fishing vessels, and nothing was known about the life of the salmon in the sea. Even now most of the information about it is derived from an examination of scales. These increase in size as the whole animal increases in size, and the continual addition of material to the periphery leads to the appearance of a series of concentric rings. When growth is slow, these rings are close together, and when it is rapid they are wide apart. The salmon, returning from the sea, has alternating bands of close-set rings and widely separated rings, and from this it is deduced that the fish has been in places where, presumably because of the cold, growth in winter is slower than growth in summer. This being so, the age of the fish can easily be read from the scales, and since, furthermore, there is a fairly good correlation between the size of the scale and the size of the fish, it is possible to tell how big the fish was at any given time in the past.

The factors which stimulate the salmon to return to fresh water are unknown, though the problem is one of both commercial and intrinsic interest, as the time of return is so variable. Some salmon, known as grilse, are back in coastal waters a year after leaving them; others may spend two, three, four, or five years in the sea. Nor is there any greater degree of regularity in the time at which the fish return from their feeding-grounds. Some, known as spring fish, for it is in the early months of the year that they appear off the coast, have evidently, from the scales, come away before the period of rapid summer growth has started. Others, known as summer fish, arrive later, and their scales show that they have been growing rapidly immediately prior to their arrival in coastal waters. It would seem that, unlike the spring fish, they have lingered to feed before the urge to migrate has sent them seeking fresh water, and some linger longer than others.

A grilse may weigh as little as 2 lb, or as much as 14 lb., though the average weight is between 4 and 6 lb. Of the other fish it is roughly true to say that they increase by an average of 10 lb. for every year spent in the sea. Ten pounds in a year is very good growth judged by freshwater standards. These figures refer mainly to Scottish fish and come from Menzies (1925), which reference may be consulted for further information about the salmon.

Arrived in coastal waters, the salmon tend to hang about in shoals until such time as there is a good flow in the rivers. They then enter fresh water and begin a leisurely journey upstream, usually, as far as can be made out, up the stream in which they themselves originated. Serious travelling does not take place until the water is fairly warm, that is April at the earliest in Scotland, and no attempt is made to negotiate obstacles as long as the temperature is low.

The most remarkable feature of this period in the life-history is that it is a time of fasting, and the fast may be of very long duration. A few fish will perhaps enter rivers in February; they will not spawn before October and probably later, and there is no doubt that some fish spend as long as ten months in fresh water. The late-comers, which do not arrive until summer is well on the wane, and often do not travel as far inland as spring fish, will have a much shorter fast, but a period of several months without anything to eat is, on the average, a regular feature in the life of the salmon. It is a phenomenon which the biologist finds difficult to explain, epecially in view of the angler's experience that, during their fast, salmon will on occasion take bait, fly, or lure.

Spawning may begin as early as mid-October and finish as late as January, though most of it is accomplished in November and December. The process was not observed in detail until Dr J. W. Jones was able not only to watch it but to film it through the plate-glass window already mentioned. The female starts with an exploration of the gravel and, after testing it at various places, begins to excavate. This she achieves by turning on her side and wafting the gravel away by flicks of her tail, The excavating is periodically interrupted while the female rests or lies in the hole, apparently testing it to see if the dimensions are right. The male meanwhile is in close attendance except when it becomes necessary to drive away another male that threatens to become a rival. Eventually the female is satisfied, after hours, sometimes days, of work. She settles in the hole and the male places himself close alongside. As she starts to extrude eggs, she opens her mouth, and the male, quivering violently, ejaculates a stream of milt over them. The female then starts another excavation upstream, and in so doing covers her eggs with a pile of gravel and small stones. The depth of each hole varies according to the size of the female, and the speed of the current. A fish weighing 12 lb. (5 kg.) may excavate a hole 12 inches (30 cm.) deep at the deepest point, and may repeat the process 12 times before shedding all her eggs.

A number of male parr do not go down to the sea but mature in fresh water. One, sometimes more, is often to be seen in attendance on a ripe female fish. Both she, and the adult mate without which she will not proceed with the successive steps of the reproductive process, will chase the parr if they see him, but most of the time he lurks under the flanks of the big fish where they cannot see him. As the female settles into the excavation for the last time, the parr inserts himself underneath and beside her down near the bottom of the pit, and as the eggs are extruded he sheds milt over them. Jones ligatured the sperm ducts of an adult salmon and then returned it to the water where it was able to induce a female to proceed to the egg-laying stage but could not fertilize them. The attendant parr did so successfully.

Jones suggests that this phenomenon may have survival value. The relatively tiny male parr is close to the eggs and his milt cannot miss them. That of the large and consequently more distant male may. It is a form of insurance that no egg shall go unfertilized, to borrow the term used by Jones (1959).

The effort exhausts the male to such an extent that he seldom survives long enough to get back to the sea. The female lives on more often, though she may have lost between one-quarter and one-third of her weight as a result of spawning. Known as a kelt, she returns to the sea and may come back to fresh water to spawn again. The whole process leaves a characteristic scar on the scales, making it possible to tell if a fish has spawned before. It is found that the percentage of females spawning a second time rarely exceeds five, but there are a few records of fish which have come back three and even four times. Menzies records the capture of a fish which had gone down to the sea originally as a two-year-old smolt and had returned to spawn for five consecutive years. He records that the oldest salmon ever caught had lived thirteen years and spawned four times.

Fishermen operating from the shores of Greenland have been catching salmon for some two centuries without attracting much attention, even though the catch increased tenfold between 1961 and 1968. In 1965 salmon were taken in drift nets off Greenland, and a tenfold increase in this catch in three years caused widespread alarm because it was evident that the salmon's feeding grounds in the sea had at last been discovered. Further depressing news for anglers was the start of similar fisheries off Norway and the Faeroes in the late sixties. The many scientists concerned with salmon have been less despondent. Pyefinch (1969) notes that though fish from many countries have been caught off Greenland, most apparently come from America; that no grilse have been taken there; and that catches in Britain have not declined as those in the sea have risen. From the commercial point of view it is critical that stocks should not fall below the level at which a constant breeding population can be maintained. This level is obviously low. A female salmon produces 650 to 700 eggs per pound of body weight (Jones, 1959), at which rate a 43-pounder (19 kg.) produces 30,000 eggs. An enormous number of the resulting progeny perish in the competition for territories, which means that only a few fish can produce all the young which a long stretch of river will hold. Therefore a heavy exploitation of the adults is possible – how heavy is the duty of the scientist to discover. Like the engineer he must multiply his answer by a safety factor of several times and, unlike the engineer, make allowance for the greater instability of the living system. Whether, when agreement has been reached on how many fish it is safe to take, the nations will agree to and be able to enforce regulations to prevent heavier exploitation is a question to which past history does not indicate an optimistic answer. Even if control is effective or, as may prove to be a possibility, the salmon's feeding grounds in the sea are so widespread that the proportion caught is never excessive, the angler is bound to feel that every salmon taken in the sea is one less that he might have caught in his river.

The salmon is an exceptionally vulnerable fish. Between the sea where it feeds and the upper reaches where it breeds there are three hazards, any one of which can eliminate it from the river in question. Of first importance in this country is the zone of polluted water which oscillates with the ebb and flow in many of our big estuaries; of first importance in Sweden, for example, is the erection of dams for hydroelectric purposes; a third factor is unrestricted fishing. A considerable fraction of the salmon stocks in the Baltic are now derived from hatcheries. Netboy (1968) takes the European countries with an Atlantic seaboard one by one, and describes the fate of the salmon in them. In spite of the query in his title (The salmon – a vanishing species?) his final conclusion is not pessimistic, though this is clearly dependent on governments being wise and well advised.

That life-histories of equal complexity and interest cannot be related for the invertebrate inhabitants of fresh water is probably because the details have not been unravelled rather than because they do not occur. It is significant that among insect groups we must turn to mosquitoes, studied so thoroughly because of their relation to disease, to find a clear-cut example of migration. *Anopheles sacharovi* in Israel leaves the neighbourhood of its breeding-places in the autumn and flies for distances of up to twenty miles to find a place to hibernate. It spends the winter in a torpid condition in some dark damp stable or cavern and then in the following spring it flies back to the breeding-place to lay its eggs. The female has been impregnated the previous year, and all the males die at the end of each season when the weather gets cold. Incidentally, one result of this migration is that villages which enjoy immunity all through the main transmission season are smitten with malaria just as it is coming to a close. Individuals of *Anopheles hyrcanus* in Bengal and Burma fly from the marshes where they breed during the winter and seek out the rice-fields as these are filled up during the monsoon. When the rains come to an end and the rice-fields begin to dry up, there is a regular flight of *A. hyrcanus* back to the marshes. But the exact nature of this mass movement is not known. One strange feature, which we learn from Major R. Senior-White, is that few mosquitoes feed while on passage.

A journey connected with reproduction is made by certain stream-living flatworms. We propose to refer to it as migration though strictly it does not conform with Heape's definition, because there is no territory where the animals are never found except on migration. *Crenobia (Planaria) alpina*, which has been mentioned several times already in other connections, first attracted the attention of Beauchamp and Ullyott because there was marked disagreement among previous workers who had observed it. Some held that it moved downstream with the current and others maintained that its normal reaction was to go against the current.

Beauchamp made a series of observations and experiments both in the field and in the laboratory. After a long period of trial and error he was able to provide the animals in captivity with conditions under which they would thrive. The main initial difficulties were due to substances dissolved out of the pipes of the water supply and from the paint which had been used inside the trough where the observations were made. This trough was several feet long and held about half an inch of water which could be kept running. An ingenious supply system was devised so that the direction of the current could be reversed at intervals. A flatworm was placed in this trough and the water supply was adjusted so that, whenever the flatworm was getting near to one end, the current was reversed and accordingly kept constant in relation to the direction of movement of the animal.

In cold springs sexually mature specimens may be found at any time of year, but generally sexual development is a winter phenomenon associated with a temperature of less than 10° C. When an individual is sexually mature, it tends to move against the current, and in his trough Beauchamp was able to keep specimens progressing steadily against the current for prodigious distances. In nature one of two things happens. The individual, toiling up the stream, may exhaust its food reserves and be unable to find sufficient forage to maintain them. When this occurs it begins to reabsorb the material which had originally been destined as a provision for the embryo. With the onset of this process the reaction of the flatworm to the current changes and thenceforth it goes downstream.

In the lower reaches it is likely to encounter a better food supply and to be able to obtain the wherewithal to mature its eggs once more. It then starts upstream again. This sequence of events may take place several times, but eventually, unless the animal is unusually unfortunate, the second of the two alternatives mentioned above comes to pass: the planarian reaches the spring-head and there it lays the cocoon containing its eggs. This done it goes downstream either until it reaches the point where the temperature is too high (see p. 150) or it has fed well enough to become sexually mature once more.

Emigration is another term used in a restricted sense by Heape. He applies it to certain mass movements of which the best known are those of the lemming and the locust. The distinguishing feature of emigration is that there is no return. It is associated with a big increase in population, which is certainly sometimes due to an abnormally high rate of reproduction. Then suddenly the greater part of the population sets off and embarks on a journey which almost always ends in disaster; apparently these mass emigrations hardly ever result in any increase in the range of the species. The whole phenomenon clearly merits a distinct name, but here we shall refer to it as 'mass emigration', leaving the term 'emigration' for use in a looser and wider sense.

Mass flights are sometimes undertaken by various insects which spend part or all of their lives in fresh water, but it is difficult to know whether they should be regarded as true mass emigrations or whether they are no more than the random wanderings which every species makes within its range. Mass flights are not uncommon among certain dragon-flies. *Libellula quadrimaculata*, a common British species, is a noted wanderer, and what may be mass emigrations of large swarms are not infrequently recorded. The species is circumboreal in distribution, occurring in both the Old World and the New. The southern limit of its range extends to Sicily and the Caucasus in Europe, and to Kashmir in Asia. The native British population is probably augmented frequently by immigration from the continent of Europe.

Swarms of another dragon-fly, *Sympetrum fonscolombii*, which has an unusual range extending from South Africa to North Europe and China, rather rarely invade Britain. Frequently they consist of males only, but on some occasions females have been observed as well. The result has been the establishment of a colony which occasionally has perpetuated itself for three or four years, but which has always died out in the end. Evidently conditions in the British Isles are not suitable for this species, and, as far as is known, it has never become established as a regular inhabitant. Its continued occurrence on the British list is due to immigration. *Sympetrum flaveolum* is taken rather more frequently than *S. fonscolombii* but it likewise maintains itself in Britain only by immigrating from continental Europe.

Records of immigrations and other mass flights refer to dragon-flies far more often than to other groups. It is pertinent to ask whether this is because dragon-flies are more powerful fliers than other insects or because they are a more thoroughly studied group and are more conspicuous on the wing. The question cannot be answered but it is germane to point out that another well-studied group, the butterflies and moths, cannot by any means be described as having great powers of flight, and yet exhibits some remarkable examples both of migrations and mass emigrations.

It is likely that among other, more obscure aquatic groups there are comparable travels, though up to now the subject has been little studied. Balfour-Browne has made some observations on the beetles invading open tubs full of water, and another group about which something is known are the water-bugs. Walton (1943) describes a long series of observations on the water-bug fauna of North Somerset, and they include some records of population changes in a single pond. On each visit Walton continued collecting till it became difficult to secure any more specimens, and then he counted and identified his catch and returned it to the pond. It is a rough and ready method of obtaining an idea of the total number of bugs present but it is sufficiently accurate for the purpose required, and a really feasible alternative technique is difficult to think of. The population began to increase in September and by early October it was about four times what it had been. Metamorphosis of nymphs into adults could not account for the rise, and it was presumably due to immigration. By mid-November the population was back at its original level. Walton suggests that in the autumn, when all the nymphs have completed their development, there must be considerable overcrowding in any pond where the year's breeding has been successful. The reaction to this is emigration and the emigrants arrive in other ponds such as the one which he had under observation. Many moved on and the composition of the population showed that different species departed at different times, one sex sometimes leaving before the other.

One may surmise, though it is no more than this, that there is much flying about of this sort but that it has been overlooked because a water-bug on the wing is not large and conspicuous, and therefore does not attract the attention even of the entomologist. It would seem that they drop into places in a haphazard way, for Walton records *Notonecta* landing on the shiny black top of a car. Another observer was taking advantage of a fine afternoon to tar the roof of an outhouse when he was surprised, and perhaps vexed, to find large Corixids landing on the tar and getting stuck in it; however he was sufficiently enlightened to call in a local entomologist, and the occurrence was put on record. It would appear, therefore, that water to an air-borne water-bug is no more than a shiny surface. But each species is confined to a narrow and well-defined range of conditions, some account of which is given on p. 160. It must, therefore, be assumed that, when a water-boatman arrives in a place where conditions are not suitable, it moves on again rather soon and ultimately finds a suitable place to live in by a process of trial and error.

If all the flying about which has been postulated really does take place, there is probably considerable mortality while it is going on. Walton was once watching water-bugs flying away from the muddy shallows of a river in the bright sunshine of an autumn afternoon, when a bat appeared and started to hawk up and down catching them. Then a sparrow-hawk swooped down and seized the bat – a most unusual food-chain, as Walton observes.

Mention of bright still autumn afternoons is found frequently in the few records which have been made. One of the present authors, walking on the Cumberland fells on a sunny windless afternoon in October, saw Corixids flying into small peat pools nearly 2,500 feet (760 m.) above sea-level. Twenty-three were caught in the space of three-quarters of an hour, and the interesting feature about the catch was that it consisted largely of a species very rare in the Lake District. If, as seems probable, the specimens had come from somewhere outside, they must have flown six miles (10 km.) at the very minimum. On another occasion when climatic conditions were similar, though the season was spring, Corixids were seen leaving a pond in considerable numbers. In travels of the type just discussed the animals take to the wing of their own volition and set out in a direction which they themselves can choose; though nothing is known about the factors by which they do orientate themselves.

Finally there is a class of travel to which Heape has given the name dispersal. The traveller is a passive participant, transported by the wind or by some other agent such as a larger animal. Obviously freshwater animals which do not have a winged adult stage cannot get from pond to pond, if there is no water connection, except by this means. Such meagre information as does exist has been discussed already in connection with the ecology of snails, on p. 147, and the remarks offered here refer only to the wind dispersal of animals which can fly. Once again a really striking instance is to be found only in the mosquito family and outside the British Isles. When the men of the Eighth Army were encamped in the desert before the battle of Alamein, mosquitoes appeared among them, apparently as a result of a strong south-easterly wind, and there was even a small amount of malaria transmitted. The surrounding land was familiar to the Army and to the Royal Air Force and was known to be desert without water or at least without suitable breeding-places for the species in question. The mosquitoes must have been transported by the wind for at least 30 miles (50 km.).

Stuart (1941) has published some information on the Chironomids breeding in rock pools at a place on the Clyde. The rocks shelve very gently down to the sea, and there is a gradation from pools which contain brackish water to pools which contain fresh water. A species which breeds in the pools nearest the sea can fly, but mating and egg-laying take place very soon after the adult has emerged – usually before either sex has used its wings. Species from the freshwater pools do not mate so soon and take to the wing before mating. Stuart suggests that, though promiscuous dispersal of the latter by the wind will cause the death of many individuals before they have reproduced, the chances that some will reach a suitable breeding-place are sufficiently high to be of benefit to the race. But the brackish-water breeder will stand a very small chance of finding a suitable place for egg-laying if it is blown away from the immediate vicinity of the sea, and its peculiar mating habits appear to be adapted to guard against such an eventuality.

Moon has described how, when the level of Windermere rises, those members of the fauna which are readily motile quickly move on to the newly inundated land. A big flood will cover grassland, and some animals, such as *Gammarus*, may be found among the grass stems. But the flat *Ecdyonurus*, adapted to spend its life clinging to stones, will not travel any farther when it comes to the edge of the stony region. A fair amount of colonization takes place within a matter of hours but it seems that the phenomenon is no more than the result of the random movements which the population is constantly making.

Berg records a general migration of the fauna of Esrom Lake into deeper water in the autumn and into shallower water in the spring. Some of the mayflies and stoneflies seek out the edge of the water immediately before emergence. Other animals seek deeper water in winter because the vegetation in the shallows dies down, and they can find the cover which they require only in the quieter regions at greater depths. But there is also a migration by animals not affected in either of these ways. The greatest number of zebra-mussels, for example, is found in deeper water in winter than in summer. There has been much discussion about reaction to temperature and to oxygen concentration but, since no clear-cut correlation has been demonstrated, there seems nothing to be gained from setting out the arguments here. This is an example of migration not directly concerned with breeding.

We must leave this aspect of freshwater natural history as we have left others – a series of isolated observations many of which are far from complete. Fish, which are large and visible to the human eye at a distance, and which moreover have a practical importance as a source of food, have been well studied for many years. The smaller animals, the invertebrates, which can scarcely be seen and certainly not identified except at close range, and which are mostly without obvious practical importance, have been little studied. Such information about them as there is shows that the subject is well worth study; but at the present stage no general conclusions or wide inferences can be drawn.

12. Stock and Crop

A preliminary to the main substance of this chapter is the topic of the relationship between the size of a fish and the attainment of maturity, for this relationship is quite unlike that found elsewhere in the animal kingdom. Most animals have a definite size when fullgrown, and variation on either side of the mean is slight. A somewhat stunted specimen may be produced by insufficient food, but, in general, if the food supply is not sufficient to enable the animal to attain the size usual for the species, it is not sufficient to support life at all. Sexual maturity is usually attained some time before full size.

But sexually mature fish of one species may be anywhere in a very wide range of sizes. For example, trout in a Highland lochan or Lake District tarn may never grow larger than ½ lb. (220 g.), and at half this size they may breed and behave in every way as normal fish. But in a good chalk stream in the south of England many of the specimens will weight 2–3 lb. (0.9–1·4 kg.), and in some of the larger lakes trout weighing 10 lb. (4·5 kg.) are not uncommon. Few other freshwater fish grow as big as this, but all show a considerable range in size.

The reason why some populations of fish are made up largely or entirely of small specimens is a matter of the greatest concern to the fisherman, and we believe that some examination of the subject will prove of interest also to naturalists who are not anglers. The late Professor Gunnar Alm, formerly director of the State Institute of Freshwater Fishery Research, described (1946) the results of many years' observations on populations of perch.

He quotes the results of the measurement of individuals in big samples of fish from thirteen different lakes. In the lake in which the fish were biggest on the average, the average length was 25·8 cm. for male and 30.9 cm. for female fish; in the lake with the smallest fish the average length for males and females was 9·9 cm. and 11·6 cm. respectively. Thus the average length in the lake with the biggest fish was nearly three times that in the lake with the smallest fish. Frequently the lengths in a large sample varied rather little on either side of the mean: in other words in any one lake all the fish tended to be large or small or medium-sized and not of many different sizes. The average size in one lake varied somewhat from year to year. The larger fish usually had a greater weight relative to length and so were better fish in every way. In general there was a relation between the average size of fish and the size of the water-body, the small fish coming from the small places.

How the rate of growth and the age of a salmon may be read from its scales has been described on p. 229. Alm found that the perch scale was less satisfactory than the salmon scale, and he was not prepared to deduce from the size of the rings the length of the fish at earlier ages. But he confirmed by extensive observations on marked fish that they indicated the age clearly, except in a few specimens of great age. The age-size investigations showed that perch in all the lakes grew at approximately the same rate for the first two years of life. After that growth might slow down almost to a standstill, or, at the other extreme, continue at the same rate. Those fish which continued to grow tended to live longer than those which did not.

Sexual maturity was found to depend on size. The male might be mature at one year old, but this was exceptional, and most mature males were two years old at least. The female was never mature at the age of one year, rarely at two, and usually not till four. Alm found that samples taken outside the breeding-season contained roughly equal numbers of the sexes, but, as might be expected from their earlier maturity, males frequently preponderated greatly in a sample of spawning fish.

Alm postulates and proceeds to examine three possible causes of poor growth: heredity, direct action of the environment, and indirect action of the environment through the food supply. Typical stunted populations were transferred to waters where growth was normally good; the transferred fish proceeded to grow well also. Alm is not prepared completely to rule out heredity, for quick-growing and slow-growing strains undoubtedly exist, but his results showed that, in general, stunted populations are not made up of fish that are constitutionally incapable of growing larger.

Investigations on the environment were carried out in a series of ponds, or tarns as Alm calls them, using the Norse word which will be familiar to north-country readers, and in a lake. This lake had been renowned for its big perch in the early years of the century, when anglers flocked to fish in it. After the 1914–18 war fishing was prohibited, and it was not resumed until the early thirties. The lake was then found to contain no big perch but a great abundance of little ones. A research team was sent to investigate, and they decided to try the effect of reducing the population. In 1937 they removed about 25,000 fish by means of traps and nets, and in subsequent years they not only continued to use these measures against the adults but attacked the population in its earliest stages. Branches were placed in the water, the female fish chose these as spawning-places, and then the removal of the branches in due course led to the destruction of a great deal of spawn. The result of this drastic reduction on the total number of fish was highly satisfactory. In 1937 most of the fish were between 14 and 16 cm. long, and the biggest taken was only 20 cm. long. Even in 1938 some effect was apparent, the most frequent length being between 17 and 18 cm. Steady improvement was maintained, and in 1945 the modal length was 21 cm., that is half as big again as it had been when the observations started, and greater than that of the longest fish taken in the first year. The conclusion is that the population had become too great after the ban had been imposed on fishing, the total food available when shared among all the fish enabled them to grow to a small size only, and a typical stunted population was the result. When the numbers were severely reduced, there was more food per head and each fish was able to attain a greater size.

The ponds, which were poor peaty tarns of the dystrophic type, contained small fish. Intensive fishing was tried, but it did not produce an increase in the average size as in the lake just mentioned. It seemed that they were incapable of supporting a population of good fish and Alm suggests three reasons: first, perch feed on zooplankton when they are small but graduate to large animals as they grow bigger. Dystrophic waters frequently have a rich zooplankton feeding on the organic matter washed in from the surrounding land, but a small population of larger invertebrates. There is therefore plenty of food for the smallest perch and many survive, but little for them when they reach the stage of requiring larger animals to eat. Secondly, the tarns are thermally stratified in summer, the hypolimnion becomes depleted of oxygen, and this restricts the volume of water in which the fish can forage. Thirdly, the tarns remain rather cold all through the season and this retards growth. Here too the conclusion is that the main reason for the stunted population is the insufficient food supply.

The relationship between growth of trout, food supply, and

hardness of the water has long been a matter for dispute. On
the whole, best growth is found in chalk and limestone areas,
and some hold that this is a direct effect of calcium on the
physiology of the fish. It could be a direct effect of some other
substance associated with calcium. Most calcereous rocks are
formed of or contain the skeletons of animals that formerly
lived in the sea. Many marine animals have the power to ex-
tract from sea-water trace elements present in minute quantity
and to store them in their tissues. Chalk and limestone rocks
may therefore be richer in these trace elements than rocks of
mineral origin. The rival school maintains that trout grow
more rapidly in hard waters because there is more to eat – it
will be recollected that in Chapter 8 attention was drawn to a
relation between the production in a lake and its calcium content, though the nature of the relationship remains unknown.
Dr M. E. Brown reports (Frost and Brown, 1967) the results of
some experiments on trout reared in Cambridge tap water in
which the concentration of calcium had been reduced in various ways. In water in which all the calcium had been replaced
by sodium, the fish died. Provided that there was some calcium
present they throve and grew at much the same rate whatever
the concentration was. However, the controls in tap water
were in a concentration of 125 parts per million of calcium
carbonate (31·25 p.p.m. Ca), for Cambridge tap water, which is
drawn from the chalk, is softened, and there is some evidence
that 150 p.p.m. calcium carbonate is a critical concentration.
The controversy is not, therefore, at an end.

If it is accepted that a stunted fish population is usually due
to an insufficient food supply, there are certain implications
which are not generally accepted by those who practise fishery
management. The first is that, since any one body of water can
obviously support a certain weight of fish and no more, the
weight of each fish will depend on the total number of fish
present. For example, if a pond can support 100 lb. of fish, it
may contain 400 fish weighing $\frac{1}{4}$ lb. each, 100 fish weighing
1 lb. each, or 25 4-pounders. This theoretical consideration is
borne out by some figures quoted by Swingle and Smith, the
American investigators whose work on adding artificial manures to fishponds is discussed in the next chapter. Three similar
ponds were dressed with fertilizer, and the authors claim that
the applications were made in such a way as to keep the food
production relatively constant. The ponds were stocked with
bluegill fingerlings of uniform size, receiving 6,400, 3,200, and

1,300 lb. respectively. At the end of the season the ponds were drained and the surviving fish counted and weighed. The average weight in the first pond was 0·8 oz., in the second 1·5, and in the third 3.8 oz., and when these figures were multiplied by the number of survivors it was found that the total weight of fish in the three ponds was 320, 300, and 306 lb. The logical deduction from this is that a fishery may sometimes benefit from a reduction of the fish population, a view which conflicts with the widely held idea that the cure for all fishing ills is to stock the water. This disagreement between common practice and research finding has caused certain scientific men to fly to an extreme and condemn *in toto* the fish hatchery and the policy of stocking. Thus Thompson (1941) writes: 'In the light of our present information it is not impertinent to inquire which helps fishing more, fish-eating birds or fish hatcheries. . . . Fish culturists seem to believe that some magic attaches to the hatchery pond and that it is the cure-all for every fishing ill. The value of the hatchery was not seriously questioned until recent years, when the results of marking experiments and of population and growth investigations began to show that it has a negligible effect on fish yield, at least in waters comparable to those of Illinois. The Illinois Department of Conservation recognizes this and has not expanded its hatcheries in recent years, nor does it plan to do so. So firmly entrenched is the old idea, however, that it will probably require years of educational effort and demonstration to convince people that it is wrong.'

No doubt money has been wasted on stocking waters where there were too many fish, and a reaction has set in. Unchecked swings generally go too far. It is plainly a matter of common sense to ascertain whether natural reproduction is replenishing losses before resorting to stocking. Generally it is, but sometimes it is not, either because spawning facilities are poor or because removal by anglers or some other predator is excessive, and then reinforcement of the natural stock is beneficial. The hatchery also has a part to play in meeting the fisherman's requirement of large fish. The largest fish in a natural population will soon be marked down by the informed anglers, and sooner or later one of them will catch it. The next largest then becomes the target, and no fish has a chance to attain a large size. Accordingly the angler who is ambitious to catch large trout, and can meet the expense, keeps his fish in the hatchery until they have attained a desirable size and then sets them free.

Incidentally, Mr I. R. H. Allan tells us of several instances where this practice has led to a population consisting almost entirely of large fish and hardly any smaller ones. It is thought that the large fish bring about this state of affairs by eating most of the smaller ones and maintain it by devouring each new generation.

In 1938 the Freshwater Biological Association undertook to investigate the biology of coarse fish at the request of the National Federation of Anglers, who provided the funds. Mr (now Archdeacon) P. H. T. Hartley (1947) was appointed to undertake the work. His investigations, as was inevitable in the time at his disposal, did not probe as deeply into the subject as some of the others mentioned in this chapter, but they merit notice because of the balanced approach and neat formulation of the problems. He writes in his introduction that, at the start of the discussions with the National Federation of Anglers, it was assumed that 'deficiencies of numbers and poor qualities of stocks were alike to be made good by the straightforward policy of restocking.... As discussions proceeded it became obvious that in making this assumption of the universal adequacy of the policy of restocking, a very great deal was taken as proven which was in fact open to question. It was seen that the fundamental problem was not the principles and practice of rearing coarse fish.'

These initial obstacles having been negotiated, Hartley set out to discover three things about each specimen of as many of the British species as were available within reach of his station in East Anglia: first what it had been eating, second how big it was, and third how old it was (from its scales). With plenty of information from many localities, and with the information culled from the continental literature, it was possible to base on the last two items conclusions about how big each species should be at a certain age, if it were not to be classed as a stunted specimen. The object was to provide a reliable yardstick by which anglers and angling bodies could judge the fish of their waters before embarking on any policy of improvement.

For further guidance fish were divided into five classes: 1, large and young; 2, large and old; 3, average; 4, small and young; 5, small and old. If the fish be found to belong to any of the first three of these categories, it may well be decided that no steps are necessary, though if fish be large, it is possible that the water might carry more fish, and stocking would improve the sport for anglers. If the fish be small and all or nearly all are young, the population is being cropped too severely either by the anglers or by some other predator. Only after further investigation can it be decided whether the best policy is to impose restrictions on the angler's bag, or to start a campaign against pike or otters or whatever else is believed to be the culprit. If the fish be small and old, and Hartley fears that in all too many British waters fish do belong to this category, then the stock must be reduced. Existing limits on total catch or minimum size may be lifted, but trapping or netting will probably be necessary too. There is also the suggestion that predators might be introduced, but this is another controversial issue and we shall return to it later.

Hartley has produced simple criteria for the use of practical men who have only their leisure hours and their own resources for their fishery and its improvement. All that is required is, first, sufficient perseverance on the part of a few individuals to learn the technique of reading scales and then to apply it to a fair sample of fish, and, secondly, sufficient co-operation from all members to ensure that a representative portion of the season's catch is made available for weighing and scale-reading.

A good example of a scientific investigation obviously involving a great deal of work, is that of K. R. Allen (1945) on a river in New Zealand. He investigated the population with a thoroughness that has not been equalled and his results are still widely quoted. For example, Frost and Brown (1967) quote him extensively in their chapter on trout and man and reproduce three of his figures. He points out that the facts about population which should be known as a basis for fishery management are:

1. The total number of fish at any one time.
2. The number in each age-group.
3. The average weight in each age-group.

The first of these categories, the total population, constantly arises as a problem in fishery work and usually proves the most difficult to solve. Allen caught many fish by netting, marked a proportion of them, and returned them to the river. Then, knowing the total number of marked fish, he calculated that this stood to the whole population as the proportion of marked fish to total in subsequent nettings. Clearly large numbers of fish must be examined if an answer worth anything is to be obtained, but Allen's work fulfilled this condition and, as he found incidentally from his marking experiments that the fish tended to stay in one part of the river, he concluded that his calculations were reasonably accurate. The total weight of fish is referred to as the stock.

4. The production, that is the total weight of fish grown in a given period. Total number of fish multiplied by average weight gives the total weight for an age-group, and the increase in the interval between two samplings is the production.
5. The crop, a term which Allen uses for the weight of fish removed by anglers.
6. The weight of fish destroyed in other ways.

When information of this kind is available the angler may ask: can the stock be made larger? Can it be cropped more heavily? Can loss due to other causes be reduced? Allen has little to suggest under the first head. Stocks of both trout and trout food are sometimes seriously reduced by floods but he fears that measures to prevent this would be beyond the means of fishing organizations. The predators were not studied in great detail, and Allen has only general remarks to make about reducing them. On increased cropping he has an important contribution to make. The high mortality of the Horokiwi trout was noticed in Chapter 8; the population is halved roughly every six months. During the first year, although many fish die, the survivors grow fast and the total weight of fish increases. It reaches a maximum in the second year. Thereafter, although the survivors continue to grow, the amount which they add to the total weight is less than that subtracted by the fish that die. Total weight declines. The population should be exploited when total weight is greatest provided the fish have had a chance to breed. On this criterion Horokiwi trout should be taken in their third year, and in the lower reaches they are. Higher up growth is slower and the 11-inch limit means that trout are not taken till their fourth year. Allen recommends a shorter limit, pointing out that the effect of the existing one is to leave a large number of three-year-olds to fall to other predators and to reduce by one-third the total weight caught by anglers.

Americans refer to the total weight of fish in a pond or lake (Allen's stock) as the carrying capacity, and they frequently measure it by the marking technique already described. But they have also obtained exact data by poisoning an entire body of water and collecting all the fish as they float to the surface. Powdered derris (5% rotenone) is applied at the rate of about 15 lb. to the acre (15 kg./ha.), and a few sticks of dynamite are often thrown in to ensure thorough mixing. The effect wears off quite soon and the locality is restocked; it is claimed that by restocking with certain species in a certain proportion the sporting properties are improved.

Thompson (1941) of the Illinois Natural History Survey gives data about the rate at which a pond may be fished. In Louisiana, latitude 30° N., a weight of fish equivalent to the carrying capacity may be removed during the course of a year without fear of overfishing. But farther north, where fish growth is slower on account of lower temperature, the proportion is lower. In Illinois, latitude 40° N., 50%, and in Wisconsin, latitude 46° N., no more than 20% of the carrying capacity should be removed in a year. These figures are quoted in order to illustrate an approach. No general principle can be advanced, because the rate of cropping by predators other than man must vary so much from water to water. The mortality in the Horokiwi has already been mentioned. A big contrast was provided by a small tarn in the Lake District which was first cleared of trout and then stocked with 500 in their first year. Over the next five years no less than 428 were caught. The tarn is too small for a population of piscivorous fish, and fish-eating birds rarely visit it. Otters were trapped in earlier years when the tarn was in private hands. Man, therefore, was the main cause of mortality and he could exploit the stock much more heavily than the anglers of the Horokiwi, who had to share theirs with eels and sundry birds.

But to pass from some of the more ambitious to some of the more humdrum aspects of American fish conservation, we find that a great deal of importance is attached to 'catch per unit effort'. Many of the fishing clubs are well organized and disciplined, and a careful record is kept of the number of hours which each member spends fishing, and the number and weight of the fish which he catches. As a check, fish are netted and the average weight of fish per haul recorded. If the amount of fish caught in the average hour on a line, or per average haul of the net, declines, overfishing is diagnosed. But, if people catch a lot of little fish, it is concluded that they are not fishing enough.

Overfishing may be counteracted by stocking, and by measures against predators; underfishing by trapping, netting, and introducing predators. This last, already mentioned as a controversial topic, is not so regarded by the Americans. Their literature contains precise instructions about the proportion of predators to pan fish (in Britain we should call them table fish) which should be introduced into a pond or lake which has been poisoned or emptied. For European waters opinion is much less definite, though Alm (1946) records improvement in the average size of perch after the introduction of pike into some lakes in Sweden.

An account of the extensive trapping and netting in Windermere will serve to emphasize the point made above that each body of water has its peculiarities, and the outcome of any steps to improve it will not always accord with forecasts based on experience elsewhere. It is probably safe to make the generalization that the simpler the community, the more accurate will be the forecast. In a tarn covering an acre or two where only trout occur, surprises are less likely than in a lake where several species interact. To make clear the purpose of the operation in Windermere a recapitulation of the relationships of the common fish will be useful. Plankton is taken by char and by small perch, but, as the former feed out in the open in deep water and the latter in shallow water, the competition between them is probably slight. The invertebrate fauna of the stones and weeds in shallow water is preyed on by moderate-sized perch and moderate-sized trout, which means most of the population of both species, and also by the eel, though this fish probably feeds more in deeper water. Small and moderate-sized fish are eaten by large perch, large trout, and pike, and of these the last is the most important; it feeds mainly on perch in summer and on char and trout in winter. Taking the perch as the central point, it competes for food to some extent but probably not seriously with the char; it competes with the trout; but it is a food for the pike and, if it were not abundant, greater numbers of more highly prized species might be devoured.

During the 'thirties, Windermere was thronged with innumerable perch of very small size; a few fishermen who knew exactly where to go could be seen returning triumphantly with 2- or 3-pounders, but for the majority half a dozen to the pound was a catch of big fish. There was reason to believe that this excess of perch had developed during the last few decades parallel with other changes in the lake which, as described earlier, had been proceeding relatively rapidly. And it seemed likely that with this rise the more highly valued trout and char had declined. Fishermen with a lifetime's experience of the lake spoke of bigger and better catches in their youth, and there was no reason to doubt their statements, although figures to substantiate them did not exist.

The work in other countries pointed to the conclusion that nothing but benefit would come of a drastic reduction in the number of perch. The survivors should be able to grow to a larger size, and there should be bigger trout; the absolute numbers of this fish would not be affected, for that depends on events in the inflowing becks, where the early stages are spent, but with fewer fish to prey on the available food supply the average size should increase. Char also might show an improvement. The outbreak of war, when any contribution to the nation's food supply became important, provided a second motive, and it was decided to take action. But how were the perch to be caught? There were no fishing restrictions to be relaxed; perch were pulled out in thousands by holidaymakers every year but the number was probably but a fraction of the total population. Windermere was much too big to poison, even had such a step been legal. Nor could predators such as pike and cormorants be encouraged, for the local fishermen would have asked pertinently how it was known that their depredations would affect perch more than the other fish. It is one thing to introduce predators into an American lake which is being stocked after draining or poisoning and where a fresh start is possible if it turns out that a mistake has been made, but another to meddle with the balance of nature in a place where the slate cannot be wiped clean if the interference does not work out according to plan.

K. R. Allen, who had recently emigrated to New Zealand, had worked on perch caught in three ways: traps, seine-nets, and rods. The last had been a useful source of large perch but was not a method that could be used for wholesale capture. The seine-net could be used only in comparatively shallow water and off coasts that were not too steep. It remained to be seen what success could be achieved with the traps. Allen describes his as a 'cylinder of wire netting with a funnel-shaped opening leading inwards at each end'. It was a trap of this type that was eventually used, though one opening was found to be enough. These traps caught nothing in the winter, which was not unexpected, for perch are in deep water during the cold part of the year, but with the coming of spring they gave results which surpassed the most optimistic hopes. As the fish came up into shallow water in April on their way to spawn, they flocked into these traps in large numbers, till sometimes it seemed there could scarcely have been room for another fish. Why they should do this was a puzzle to the investigators at first. It was thought that curiosity might impel the fish to examine the bright objects, for the traps of new galvanized wire-netting were conspicuous. But then traps painted different colours, some black, some white, and some red, were set in the lake. To the general surprise the black traps consistently caught more than any of the others. Later it was noticed that the traps were festooned with spawn, and it seems probable that the female fish chooses objects of this type on the bottom of the lake as a place to lay her eggs. It will be remembered that the Swedes knew they could reduce a perch population by putting branches in the water and removing them when they had been covered with spawn.

The traps are lowered to the bottom of the lake, each attached to a glass ball or a cork float. A series at different depths indicate at what depth the greatest catch is to be made, and in the rest of the lake all the traps can be set at this depth. As the fish advance into shallower water, the traps are brought up nearer the shore. For some six weeks, from late April to early June, trapping proceeds hectically, then spawning is over, the fish disperse in the shallow water, and the numbers which enter traps is so small that fishing is not worthwhile. In the early years traps were emptied thrice a week during the season, but latterly once a week was found sufficient.

Sometimes a pike enters one of the traps and, although the entrance is only three inches wide, specimens of up to 4 lb. (1.8 kg.) in weight have managed to squeeze through. The perch population inside is quickly reduced by such an intruder and no more will enter as long as it is there.

1940 was a year of trial and experiment. Results were satisfactory and it became evident that here was a method whereby large numbers of perch could be removed from the lake. Two problems remained: first, what was to be done with the fish, and, second, where was the labour to come from. Clearly the fish had no value on the fishmonger's slab, for many weighed no more than 1 oz. (28 g.) A lake perch is not by any means to be despised as a table fish, but it requires skinning and it has a great many bones. The housewife could not be expected to skin enough one-ounce perch to make even a war-time meal, and few people have sufficient time or patience to pick the flesh from so tiny a skeleton. The solution came when it was discovered that the perch could be tinned. Violent disturbance in a rotary mixer was found to rub off nearly all the scales, and cooking under high pressure softened the bones as it does those of sardines or salmon. Moreover, the perch-trapping season in the spring fell conveniently at a time when sea-fish, such as sardines and sprats, were not available for canning, so Mr Banks, head of a fish-canning firm in Leeds, promised to take all the perch which could be supplied to him. He canned them in oil or war-time tomato juice and put them on the market as 'Perchines'.

The Association's staff could provide some of the necessary labour but a great deal more had to be found somewhere. The co-operation of local residents interested in the lake and its fishery was sought, and there was a generous response, almost every one who owned a rowing-boat, and many who did not, volunteering to give assistance. The lake was divided into areas, each based on some convenient landing-place, and within each area traps were allocated to people who lived or kept a boat conveniently near, usually 20–30 to each man. On the appointed day, in the evening, for many of the helpers had businesses to run or jobs to work at during the daytime, the lifting of the traps would begin. Each man emptied the traps of his beat into a box, made a record of the total weight, and then took the catch to the landing-point, leaving the traps in position once more. Then, late at night, a lorry collected the catch from each landing-place and took it away to Leeds.

In 1941 trapping started in earnest, though only in the north basin of Windermere. The season lasted ten weeks, and the catch was 25½ tons of perch, 2½ cwt. of eels, and 2 cwt. of pike. In 1942 both basins of the lake were fished, and the total catch increased somewhat, amounting to 30½ tons of perch, 5 cwt. of eels and 2½ cwt. of pike. Thenceforward both basins of the lake were fished but the catch began to decline, being 13½ tons of perch in 1943 and 6 tons in 1944. Since then it has been relatively constant at about 5 tons per year.

It is interesting that the catches stand in just the same relation if they are expressed as the average weight of fish per trap in the six best weeks of the season. The figures for the north basin during the first four years are 21.1, 11·3, 6·0, and 2·9 lb. per trap, and thereafter the amount remains roughly the same.

These figures show a drop to about half the previous catch each year up to 1944, just as those for total catch did. The south basin figures are similar, but the average is a little higher each year, which is interesting since other studies have indicated that the south basin of Windermere, which receives the sewage from Bowness and Windermere town and the drainage from the fairly thickly populated Esthwaite valley, is more productive than the north.

Obviously this drop in catch was related to a reduction in the total population; but just how is not known, for there has been no information about the total population of perch in the lake. Total catch, and catch per trap, after dropping steadily became constant at about one-sixth or one-seventh of the original total. Possibly this is a measure of the reduction of the whole population, but until certain aspects of perch behaviour have been investigated further, a conclusion would be premature. When the population of a certain area is thinned by netting or trapping, infiltration from adjoining areas probably brings about uniform fish density in a fairly short time. But this may not continue indefinitely and there may be a density at which there is little tendency to invade adjoining areas, whether these be sparsely inhabited or not. The traps evidently attract fish, but from how far they attract them or to what extent the principle just enunciated may operate is not known. Other objects may be more attractive than the traps, but whether this be so, what the objects are if it is, and how frequently they occur on the bottom of the lake are all unanswered questions. These are some of the unknown factors which make a simple relationship between catch and total population unlikely.

After the 1947 season the co-operative effort was brought to an end, but since then traps have been set each year in order to obtain a sample for analysis. As had been expected, the average weight of perch has increased, and is now about three times as great as it was when trapping started. The surprise has been that the population had remained at a steady level and has not started to return to what it was at the outbreak of war either gradually or by leaps. The latter is what might have been expected, for the success of a class varies greatly from year to year. It does not apparently depend on the number of eggs laid but is connected with warm summers. That a successful year-class is successful in all the lakes is evidence that the factor is climatic, for it is difficult to see what else could be common to the various lakes.

A discussion of why the numbers of perch have remained low must begin in the war years. Once the reduction had started, it became obvious that pike must be thinned too, lest they compensate for the scarcity of perch by eating more char and trout. It was found that they could be caught in gill nets in winter, but it was necessary to use nets with a large mesh in order to avoid catching trout and char. In fact a few monster trout, weighing up to 10 lb. (4.5 kg.), are captured, but as fish of this size are not caught by the angler and as their feeding habits are similar to those of the pike, their destruction is beneficial. The mesh is of such a size that all one- and two-year-old and most three-year-old pike pass through. The oldest fish caught in the earlier years was seventeen, and some 20% were nine years old or more, but as the years passed fish of this age became very scarce in the nets and their numbers had evidently been drastically reduced. The total population, too, fell to begin with, but recovered and indeed surpassed the figure for 1944, the year in which netting started. The likely explanation is that the disappearance of the old large pike has led to a better survival of younger fish, which now thrive, their numbers unchecked by cannibalism, until they are big enough to be caught in the nets.

The failure of the perch to regain their original numbers has been attributed to the predation on them by the increased numbers of young pike.

Perch are undoubtedly bigger than they were, and in this both the forecast and the experiment were successful. Catches in gill and seine nets of char are now larger than they were, and anglers agree that there are now more and bigger char than when interference started. Here, too, success has been achieved. Whether trout are bigger and more numerous is less certain. Frost and Brown (1967) state that there is no scientific evidence but that anglers believe that they are.

It appears that at the moment equilibrium is being maintained by the pike-netting in winter. Not all anglers are pleased; there are some who prefer to catch large numbers of tiny perch rather than fewer larger ones, and others who deplore the scarcity of large pike. The question that fascinates the scientists involved is whether the equilibrium is stable or not. At the time of writing (Feb. 1970) the progeny originating in two good summers (1968, 1969) is about to swim into the scientists' ken and the numbers of trout have probably been reduced by the salmon disease UDN. Like those of many another scientific project, the importance of the results increases with the number of years during which observations are made.

13. Fishponds and Manuring

Three topics, all of which have been discussed already, demand further notice as an introduction to this theme. The first is fish growth. The growth of fish will not be good unless they are getting enough to eat, but, when the food supply is sufficient, other factors which affect growth come into play. The maximum rate of growth is different in different kinds of fish; thus, according to the German fish cultivators, a carp will increase its weight by about 12 oz. (340 g.) during its second growing-season, whereas a trout will not put on quite as much as half that amount. The growth-rate may vary within one species. Commercial cultivators have long been selecting the fastest growers as their breeding stock, and in recent years there has been interest in this subject in sporting circles. Little has been done with European species, but the Americans have been active. Their work is reviewed by Hickling (1962) who reports that persistent selection over periods of twenty years and more has produced strains that grow more than twice as fast as the ancestral stock. Anglers will, however, be disappointed to learn that, when they were released among wild fish, they disappeared after a few years.

Temperature is of paramount importance to fish growth. Five hundred kg. of carp flesh per hectare (kg./ha. and lb./acre are almost the same) is very good production in Europe, but the Israelis, in their warmer climate, obtain an average yield of 2,200 kg./ha. and have recorded as much as 3,500. Hickling points out that such figures, though useful for comparison, are not scientifically accurate, because, if the fish are fed, as they usually are, the weight should be calculated not per unit area of pond but per unit area of the land on which the food was grown as well.

The optimum temperature for salmonids lies below 20° C., and between 20° and 30° C. they die. Carp on the other hand flourish between 30° C. and 40° C., and *Tilapia*, a tropical fish, has a higher optimum still. As has been pointed out mammals have an advantage over fish in cold climates because they maintain a constant temperature and the functions of the body are not slowed down. Energy has to be diverted from growth to keep up temperature, but growth does proceed. In tropical temperatures in contrast the warm-blooded animal has to use energy to get rid of heat, whereas the cold-blooded animal has not. For this reason fish are important producers of protein in tropical countries.

The final preliminary point is the place in the food-chain of the fish which is to be cultivated. Each link yields only about a tenth of the amount of living matter produced by the link before it. If, therefore, a piece of water contains a plant-eating species, it will produce ten times more fish than if stocked with a carnivorous species. The weight of fish-eating fish, such as the pike, will be ten times smaller again. The figure ten is an approximation which has been discussed in Chapter 9 (p. 186).

Certain species of *Tilapia* are the only truly herbivorous fish which up to the present have been found suitable for cultivation in ponds, and this is a further reason for the importance of the practice in the tropics.

It must be stressed at the outset that a consideration of fish cultivation cannot be divorced from finance and even from politics. The practice in any one country must be viewed against the national background before the scientific aspects are evaluated. The cultivation of freshwater fish has been widespread in eastern Europe for a long time, presumably to meet a demand that could not be met from the sea as in countries with a larger seaboard. Our scientific knowledge of it we owe largely to the Germans, who published many papers between the wars. To begin with it was a necessity in view of the allied blockade and later it was probably officially encouraged as a measure to ease the siege conditions which might be experienced if a second war started. The German approach was coloured by the need to make the most of every resource. In America, where extensive exploration of this field has been more recent, such an approach has not been necessary. Fish-farming has been more of an ancillary, and a useful way of putting to a secondary use shallow lakes dammed to supply water in dry parts of the country. In Israel carp are cultivated in regions where the water is brackish, for carp can tolerate these conditions and other crops cannot. It was a way to use land that would otherwise have lain derelict and in fact it is not always easy to sell the fish produced. In Africa, India, and the Far East fish cultivation is important because it is a particularly economical way of producing protein. Here too, careful husbanding of resources is essential. Dr G. A. Prowse, Director of the Fish Breeding Research Station at Malacca, has related how, when he imported some new fish-meal pellets to experiment with, the natives quickly discovered that a tasty soup could be made from them. An extra link in the food-chain was evidently unnecessary and the experiment was not proceeded with.

In Britain it is now a matter of centuries since freshwater fish have been cultivated for any but sporting purposes, and the stew-pond of the monastery and the manor has long since fallen into disuse. Presumably the rise of the sea-fishing industry and the improvement of communications between fishing-ports and inland markets have combined to bring this about. When war broke out in 1939, thoughts turned to freshwater fish as a possible source of food, and accordingly Dr C. H. Mortimer, who had a good knowledge of German, undertook to read and make a summary of all the papers on fish cultivation available. In the event there was no further progress; local effort was directed to the perch trapping described in the last chapter, and many of the shallow lakes in the south and east of England, which might have been suitable, were drained because at night they provided landmarks for hostile aircraft. However, after the war Mortimer's summaries were brought up to date and published (Mortimer and Hickling, 1954). Dr C. F. Hickling has since written a definitive work on the subject (Hickling, 1962) devoting much of it to the work that has been done in the tropics. A later work (Hickling, 1968) is much shorter.

There have been trout farms to supply fish for sporting purposes for many years in Britain, and latterly a number have been established to provide fish for the table. These commercial ventures naturally do not publish their methods for their competitors to copy.

As already stated the centre of European fish cultivation lay to the east and much of our knowledge about it is due to the Germans. The standard work is that of Schäperclaus (1933) and it is now available in English. It is one of those thorough books for which the Germans are celebrated, and from it the reader may even glean instruction about how to keep his accounts.

Most of the European work has been done with three fish, the carp because of its rapid growth, the trout because of its high value as a table fish, and the tench because of its hardiness and ability to do well under conditions which prove unfavourable for the other two.

Schäperclaus gives some figures about the production of fish which may be expected in ponds of different types. In a typical moorland peaty fishpond the yield of fish in a year is put at between 25 and 50 lb. per acre; in a lowland pond with a hard water supply and a rich silted bottom 200–400 lb. of fish per acre may be grown in a year. These figures are similar to those of the farmer, who reckons that pasture will produce, according to its quality, between 5 and 300 lb. of meat per acre per annum. The data just quoted are based on observations on ponds stocked with one-year-old carp. The trout cultivator cannot hope to do as well, for, according to Schäperclaus, though the average yearling carp and yearling trout are both of approximately the same weight, about 2 oz. (56 g.), the carp will weigh some 14 oz. (400 g.) at the end of its second growing-season, and the trout only 7 oz. (200 g.).

The problem which interested the Germans was how to obtain 400 lb. of fish from a pond which, left to itself, would produce only 25 lb. It is significant that Germany had not only the biggest organization for the study of purely practical aspects of freshwater fish production, but also, at least in Europe, the biggest organization of theoretical freshwater biologists. It is a question of providing more food for the fish, which can be done in two ways: by adding food; or by tackling the base of the food chain and producing more natural food. Both methods are frequently used in conjunction. The second is the one with scientific interest and the first is predominantly commercial, the main consideration being what is available cheaply. Meal made of soya and lupin and many other plants has been used for carp and any surplus animal tissue from abattoir or fishery for trout.

From what has been said above in Chapter 2, it will be clear that the nature of the pond bottom is of paramount importance. A good bottom is a mixture of organic matter derived from the dead bodies of animals and plants and of inorganic matter in a fine state of division, that is silt, of which the importance was stressed on p. 119. On such a bottom decomposition proceeds more rapidly than on other types, and another characteristic of importance is that it absorbs chemical substances added to the water to increase the growth of living matter. They return to solution slowly and gradually. The other two types into which bottoms, broadly speaking, can be classified, are the peaty and the inorganic. An acid peaty soil tends to bind nutrient salts so tightly that they are removed from circulation whereas an inorganic soil does not bind them at all, and most are washed out of the pond before any plants have been able to make use of them.

The peaty bottom consists of plant remains which have undergone little or no decomposition. The absence of calcium is commonly one of the reasons why decay has not taken place, so that the addition of lime will usually bring it about, and produce a more satisfactory substratum. It is stated that 7–8 cwt. per acre is sufficient for a bottom which is just acid, but the dose must be increased with increasing acidity; the most extreme conditions, which will be a pH under 4, should receive 30–60 cwt. per acre of calcium carbonate or half that amount of lime.

Inorganic bottoms consist chiefly of clay, sand, or gravel, and they may be improved by an addition of stable manure or sewage sludge.

European practitioners are agreed that it is beneficial to allow the pond to dry out for a period of a few weeks each year; it is, of course, usually done in winter when the fish are not growing and are in special winter ponds where they can be kept closely crowded together. What processes take place is not known, but it is believed that exposure to the air causes oxidation of the mud to a depth greater than occurs under water, and that this has a beneficial effect on production when the pond is refilled. It is probable that in cold weather the soil will retain enough moisture to keep many of the mud-dwelling animals alive. There is disagreement, on the other hand, about the advantages of the practice of leaving a pond dry for a whole year and planting leguminous crops on the bottom. Accumulation of nitrogen from the nitrogen-fixing bacteria associated with Leguminosae is one obvious advantage of this technique, but it may, perhaps, not compensate for the destruction of the bottom fauna.

The European workers emphasize that the best results from the application of the various manures which they have tried will not be obtained unless the lime requirements of the water have been met. They estimate that the lowest concentration for really good production is 65 parts per million of calcium carbonate. This is a little more than 20 parts of calcium per million, a figure which, as noted in Chapter 8 (p. 156), appears to be significant in the ecology of aquatic animals.

Of the various manurial substances which have been tried,

phosphate has given the most satisfactory results, and increases in productivity of the order of 100% have been claimed. It may be obtained as super-phosphate, basic slag, bone meal, or di-calcium phosphate; the last three contain a certain amount of calcium, for which allowance must be made when the lime requirements of the pond are being calculated. Between 1 and 1½ cwt. of phosphate per acre is the recommended quantity, and it should be applied in two doses, the first in May or June, about a fortnight after liming, where this has been necessary, and the second in July or August. A larger quantity will produce a greater yield of fish, but the increase will not be proportional to the extra amount of phosphate applied.

The European workers find that the addition of half as much potassium to the phosphate is beneficial, but their results with nitrogen in the form of nitrates or ammonium salts have been conflicting.

Swingle and Smith (1941) of the Agricultural Experimental Station, Alabama, are perhaps the two most familiar names in American fish-culture. These workers set out to find what substances were required to produce a maximum growth of phytoplankton, and discovered that the most important elements are nitrogen, phosphorus and potassium mixed together in a certain proportion. These substances were available in a commercial artificial manure known as '6-8-4 (N-P-K)', a title which denotes a content of 6% available nitrogen, 8% phosphoric acid, and 4% potash, and could most easily be obtained in this form. The proportion of nitrogen in this mixture is too low to promote the maximum growth of phytoplankton and so 10 lb. of sodium nitrate should be added to every 100 lb. of it.

This manure is applied at intervals at the rate of 100 lb. to the acre. The first dose is put on in the spring when vegetation first shows signs of growth. The result is such a rich growth of phytoplankton that the water assumes a pea-soup appearance, and the bottom is not visible at a depth of 18 inches. As soon as it is visible at this depth, another dose is applied and this criterion is used to denote when further applications are required all through the summer. The effect of maintaining such a dense crop of phytoplankton is to deprive of light and so eliminate all rooted vegetation, an end which the Americans believe to be desirable. It would certainly not be desirable in a trout pond, for the cover provided by rooted vegetation is essential for the survival of the animals on which trout feed. We are surprised at our failure to find in the literature any detailed account of the changes that follow any particular treatment. There is information about what becomes more abundant, but no account of how the abundance of one organism affects the abundance of others. The literature is of course very large, and as this kind of work has a commercial significance it is often published in the language of the country. The searcher is therefore even less confident than usual that he has not missed anything important.

Owners of ponds in Britain have expressed interest in the possibility of improving the yield of trout by adding fertilizers, but experience is still meagre.

Frost and Brown (1967) refer to the fertilization of a Lake District tarn and conclude that 'the improvement in growth of the trout has been variable and not very impressive'. In this work too it was impossible to control the population and difficult to ascertain its size, without a knowledge of which conclusions about the effect of any other factor affecting rate of growth are speculative. It may well be that the flow through ponds of this type is too great and any nutrients added are washed out too quickly. It was found that a tarn near Three Dubs Tarn, the one referred to by Frost and Brown, received drainage from an area thirty times as great as its own. Its mean depth was probably about thirty inches and therefore one inch of rain would fill the tarn. One inch of rain in a day is commonplace in the Lake District, and if it fell on waterlogged soil on a windy day, it could replace all the water in the tarn. As the conditions mentioned are fulfilled fairly often, a complete change of all the water within twenty-four hours is far from being a rare event.

Swingle and Smith estimate that untreated Alabama fishponds, according to their nature, produce as little as 40 or as much as 200 lb. of fish per acre per annum. This range is similar to that found in European waters; the fish are quite different on the two continents, but all are 'coarse' in the terminology of the fisherman. Addition of chemical manures to Alabama ponds leads to a production of 500–600 lb. of fish per acre per annum. This is rather more than the production in European ponds, but, then, the rate of manuring is considerably greater. When the applications are made according to the transparency of the water, between eight and fourteen will be necessary during the course of the season, the actual number depending on the natural salt content of the water. One acre will receive, therefore, anything from 800 to 1,400 lb. during the year. A recommended German dose is $1\frac{1}{2}$ cwt. of phosphate and $\frac{3}{4}$ cwt. of potassium, which is a little under 250 lb. per acre per annum. The American dosage is, therefore, three to six times as great as the European.

A slightly different mixture, '4-8-10 (N-P-K)', has been used in Canada in the Quebec region, where it is estimated that the average natural yield of trout is only about 6 lb. per acre. The manure was added to three experimental ponds in different amounts, and the yield in three months was 65 lb. of fish with a dose of 64 lb. to the acre, 70 lb. of fish with three times this amount, and 121 lb. with five times this amount. The irregular results are attributed to the fact that the amount of water flowing through each pond was not the same. It seems likely also that the German finding that an increase above a certain dose does not lead to a commensurate increase in the production of fish may explain these results in part at least. The work is only preliminary but it does give a pointer to those who are interested in the production of trout.

In America less importance is attached to liming ponds deficient in calcium, and the only recommendation is that, for waters which are neutral or acid, 15 lb. of powdered limestone should be added to every 100 lb. of 6-8-4 (N-P-K) mixture. It is also recommended that ammonium sulphate should be avoided as a source of nitrogen in acid waters as it tends to produce sulphuric acid. It should be replaced by a nitrate salt.

Organic manures may also be used for increasing production in fish-ponds, and a variety of substances, such as farmyard manure, certain waste vegetable products, and even the weeds dragged from another pond have been applied. The Americans claim good results with cotton-seed meal, and state that the effect is nearly doubled if superphosphate be mixed with it. The addition of substances of this nature to a fish-pond must be carried out with caution, and frequent small applications are necessary; too large a quantity may, in the process of decomposition, use up so much of the oxygen dissolved in the water that the fish will die.

The organic matter in sewage has been used for many years in Europe to increase fish production, but the practice has never caught on in this country, and the health authorities react unfavourably to any suggestion that it should be tried; they remain unconvinced by the continental claim that there is no risk of infection with sewage-borne diseases. In Europe fish are kept for a week or two in clean water, between their removal from the sewage ponds and their despatch to market, and it is maintained that during this period they become free of all organisms which cause disease in man.

One of the best-known installations is at Munich, which possesses three features essential for the successful application of this practice. First there was available an area of ground which was suitable as a site for the ponds and was of no great value for any other purpose; secondly, a clean river is available to dilute the sewage liquor with well oxygenated water; and, thirdly, the sewage effluent is not contaminated with any industrial waste-product toxic to fish. The sewage effluent is passed through the usual screens and sedimentation tanks and then, having been diluted with between four and ten times its own volume of well oxygenated water, it is run into the fish-ponds.

One acre of fish-pond six feet deep will purify the domestic sewage from about 1,000 people, and the production of carp in this volume of water is of the order of 500 lb. per annum according to the published figures. In the Munich fish-ponds rainbow trout have been cultivated successfully, in addition to carp and tench.

In hot still weather there may be a risk that decomposition will deplete the ponds of oxygen, and it may be necessary to cut off the supply of sewage liquor for a period of several days. There must, therefore, be an alternative method of sewage disposal at a time when successful disposal is at its most difficult, and this is a fourth condition which has to be fulfilled before the purification of sewage in fish-ponds is feasible.

One further obvious but nevertheless sometimes overlooked fact about fish growth, which requires notice before this discussion comes to a close, is that maximum production cannot be obtained if the initial stock is too small. In fact, the size of the best initial population has been the subject of a large amount of research. Rather detailed instructions are available to continental fish-farmers for stocking ponds which can be drained, and the following is an example. The stock for an acre of pond, of which the estimated production in a year is 100 lb., would be: yearlings, 45 carp, 50 tench, and 35 trout; two-year-olds, 20 each of the same three species. It is estimated that deaths among the yearlings would carry off 20% of the tench, 10% of the carp, and between 2% and 5% of the trout, but among the two-year-olds deaths would only amount to between 2% and 5% of all the species.

41. Impure Water

Man has probably used streams and rivers to dispose of his refuse ever since he started to live in settlements. The practice, however, did not lead to really unpleasant results until he began to forgather in large towns and to dump into the nearest river not only his domestic waste material, but also that which emanated from the factories in which he was employed. Fitter (1945) records that the last salmon was taken from the River Thames in 1833 and, to judge from other rivers, this indicates that bad pollution had been affecting the river for several decades previously. In 1827 an anonymous pamphlet drew attention to the disgusting state of affairs whereby the Thames was used both as a sewer and as a water supply. Conditions over the rest of the country were no better though perhaps they did not get so bad quite so soon. The first step in any sort of retrenchment was the Public Health Act of 1875 and the Rivers Pollution Prevention Act of 1876, but neither achieved its full object and the battle for pure rivers is still raging today.

Both fish and human disease were mentioned in the opening paragraph, and pollution has a slightly different meaning according to whether it is seen by the angler or the hygiene officer, a difference which sometimes leads to confusion. If crude sewage is released into the water, the risk of infection by the organisms causing certain diseases renders it unfit to drink, and it is polluted. A certain amount of sewage, however, is beneficial to fisheries, as was described in the preceding chapter, and it is not until sewage is present in excess, that the angler considers a stream polluted. An effluent that has been purified enriches a water with phosphate, nitrate, and other salts, and this is another cause of change which may be undesirable if it goes too far. Anything that affects fish, or the animals which they feed on, adversely is pollution. At this point we pass from the biological to the legal field, and the safest course is to quote a lawyer: 'At common law the riparian owner of land has a natural right to have the water flowing past or through his land in its natural state of purity' (Gregory, 1967).

The commonest form of pollution is that caused by sewage.

The important constituent is organic matter, which consists of carbon and hydrogen linked together in chains, attached to which are nitrogen and certain other elements, notably sulphur and phosphorus. As soon as it is released into water, this organic matter is readily attacked by bacteria, which find in it a rich source of energy. The result of their activities is the breakdown of the complex organic compounds into simple inorganic substances such as water (H_2O), carbon dioxide (CO_2), nitrate (NO_3), sulphate (SO_4), and phosphate (P_2O_5). All these compounds, it will be noted, contain oxygen. When a sewage effluent is turned into a river the immediate result is, therefore, a rapid using up of the oxygen dissolved in the water. If there is still some organic matter left when all the oxygen has been used up, decomposition still goes on but now it is caused by a different set of bacteria which do not require free oxygen. Typical end-products are methane or marsh gas (CH_4), ammonia (NH_3), and sulphuretted hydrogen (H_2S). Some of these compounds are poisonous.

The speed of a biological activity increases with increasing temperature. If sewage is discharged into warm water, breakdown takes place within a short space of time and therefore within a short space of river. In colder water the decomposition will be extended over a greater length of river, the sewage will get mixed with a larger volume of water, and the chances of the oxygen supply being sufficient are improved.

If the amount of sewage is small and the quantity of water large, the reduction in the concentration of oxygen will not be great enough to harm any organism. But a sewage effluent always brings in a certain amount of suspended matter, and this may settle in quiet parts of the river and form banks of sludge. If these cover gravel stretches where trout are wont to spawn, recruitment will be affected adversely, and the new fauna may provide less food for fish than the one replaced.

More sewage and less water will lead to a state of affairs where the oxygen is reduced to a value at which some animals and plants will die. A further increase in the relative amount of sewage will lead to the serious situation where oxygen is totally depleted and compounds which are toxic to life are produced.

The effect of an effluent must be considered not in relation to the average flow of a river but in relation to minimum flow and maximum temperature. The whole question was investigated very thoroughly by the Royal Commission on Sewage

Disposal which, appointed in 1898, issued its final report, the ninth, in 1915. The Commission's recommendations were that, where there is a dilution of 1 volume of sewage to 500 volumes of water, crude sewage can be discharged without harmful result. Where the dilution is between 300 and 500 the effluent should conform to a standard which can, in effect, be attained by screening off the larger solid matter and allowing fine particles in suspension to settle. When the dilution is between 150 and 300, an effluent of the standard which they recommend can be obtained by some chemical precipitation in addition to screening and sedimentation. But if there is less than 150 volumes of water to 1 volume of sewage, then the sewage should be subjected to complete treatment.

Complete treatment is usually necessary, and there are many ways in which it can be done. The subject is a big one and its literature extensive, but an excellent summary of modern practice is given by Southgate (1969). There are five main methods. Septic tanks are now used only for small communities and single houses in the country. The sewage is run into an excavation in the ground and there it decomposes without oxygen. The liquor percolates away into the soil and the solid sediment is removed at intervals. In some places this method cannot be bettered. If there is a stretch of permeable soil between the tank and the nearest water-course, most of the products of decomposition will be absorbed by the soil, and the water will not be enriched. This is often a desirable aim for, though some enrichment is good, too much may lead to undesirable changes in the animal and plant communities. This is particularly true if the stream runs into a lake, in which enrichment can lead to deoxygenation of the hypolimnion. Septic tanks are not suitable in rocky areas or anywhere where the effluent runs straight into the nearest stream or river, because the products of decomposition without oxygen are toxic.

In all the other methods there is some preliminary treatment in which solid matter is strained off and allowed to settle. The liquor is then passed on for further treatment. It may be subjected to precipitation by chemicals; this is a popular method in America, though it is not widely practised in this country except where certain trade effluents make it necessary.

An old-fashioned method, now almost obsolete, is to run the sewage on to the land and allow decomposition by bacteria and other biological agencies to take place there. Provided that there is plenty of suitable land available – it should be flat with a porous soil – that the location is reasonably remote so that nuisance is not caused, and that there is no danger of polluting water supplies through fissures, this remains a satisfactory method of purification. It is, however, an uneconomical use of land. Such places are known as sewage farms, but the gain in fertility of the fields that can be cultivated does not compensate for the loss of those fields which must be left under irrigation during the growing season.

The most popular method today is the sprinkling or trickling filter, in which the organic matter is also broken down by biological agencies, but in a much smaller compass. A sprinkling filter with its squat cylindrical chamber and its rotating arms is, for some reason, a common object of the countryside near railway lines. The liquor, having been distributed by the rotating arms, trickles down through a bed of broken stone, coke, or clinker. The very large surface exposed by these materials is colonized by a complex film of living slime consisting of bacteria and protozoa with higher animals, chiefly worms and the larvae of midges and moth-flies, dwelling in it. The component organisms absorb organic matter and break it down, and the final effluent, after passing through a last sedimentation tank, may be pure enough to drink.

The most modern treatment is the activated sludge or bio-aeration process. It depends on the presence of certain bacteria which cause the organic matter to precipitate as a sludge or floc, on which they, protozoa, and other organisms cluster. The sewage liquor is agitated by compressed air or some other means at the beginning of the process, and then the floc is left to settle. A small amount of it is mixed with freshly introduced liquor to initiate the process of precipitation, and the rest is removed to drying beds.

Besides sewage, a number of trade wastes are commonly discharged into rivers, often with disastrous effect. Beet-sugar factories, places where milk is handled, breweries, distilleries, slaughter-houses, and flax-retting plants all produce effluents which are essentially similar to sewage, since the effective component is organic matter. The concentration, however, is usually greater. They can be treated in the same way as sewage and can be admitted to the local authority's sewage works without causing any difficulty, though allowance may have to be made for the higher concentration.

Hynes (1960) distinguishes seven types of pollution, the first of which is that by organic matter just described:

2. Poisons. Copper, zinc, and lead come from various industries and mines, chromium salts from the leather industry and from electro-plating, phenols and cyanides from chemical industries and gas-works, and tar acids and ammonia from gas-works. Insecticides have recently provided many new problems.

3. Inorganic reducing agents, notably sulphites and sulphides.

4. Oil, which kills any animal that comes to the surface for air and which hinders the oxygenation of water.

5. Inert suspensions of finely divided matter enter the water from places where china clay is extracted. The particles settle on the bottom and form a layer in which there is no food. Coal-dust, sawdust, and fragments from paper mills can reduce the population of a stream in the same way; though the last two are organic and may use oxygen by decomposition.

6. Hot water flows out of many power-stations and other places where cooling is required. Alabaster (1965) reports that power stations may heat water by as much as $10°$ C., which is likely to take the maximum temperature of a British river to just over $30°$ C. This is not lethal to many animals, though in the present state of knowledge it is not possible to list those species which can tolerate this temperature and those which cannot. Since fish avoid unfavourable conditions, and, since warm water tends to float on the colder water below, Alabaster concludes that heated effluents are not doing much harm. However, as noted above, decomposition is quicker at higher temperature, and a heated effluent entering a polluted river may cause the oxygen concentration to fall to a critical level that it would not reach at the slower rate of decomposition in the normal temperature of the river. Moreover Hynes points out that the water is sometimes chlorinated, if it is polluted, and that the chlorine forms toxic compounds with thiocyanates, if any are present in the water.

7. Non-toxic salts may produce a concentration that is fatal to freshwater organisms, but rarely do so in Britain where there is little mining for salt.

Hynes devotes a chapter to 'other human influences on natural waters' and in it describes certain activities which, though not commonly regarded as pollution, might render the perpetrators liable to a legal injunction. Deforestation increases the risk of harmful spates in streams draining the region cleared. Weirs, well known as a menace to migratory fish, may retard the recolonization of a stretch of river that has been depopulated. Canalization of rivers to improve drainage can devastate the fauna, and an example which Hynes quotes from his own experience is sufficiently striking to justify repetition here. A small south-country trout stream meandered across a clay plain through pools, through scattered weed beds growing on shoals of silt, and over gravel riffles. 'Improvement' with mechanical shovels, left it a straight canal in a bed of stiff clay. After the operation Hynes could find 25 specimens in ten minutes where he had taken 254 before it. He believes that it will be years before gravel accumulates once more, for there is little stone in the drainage area, and that, until it does, the invertebrate fauna will remain sparse.

The big rise and fall of the level of a lake that has been dammed to control flooding or to provide water may eliminate many of the inhabitants of the shallow water. *Coregonus*, which lays its eggs in shallow water, has disappeared from many a Scandinavian lake after it has been dammed to provide power for electricity. Hynes reports that in Llyn Tegid (Lake Bala), where it is known as the gwyniad, this fish is at risk, but he is writing too soon after the damming to be certain of the outcome.

It is common in dry countries to pour fatty alcohols onto reservoirs because they form a film on the surface that impedes evaporation, and the process has been used in Britain. The film apparently does not slow down the diffusion of gases in and out of the water, but it does hinder the emergence of insects.

The list of toxic substances given here is long, though far from complete. Many of the substances, however, only rarely find their way into our waters in lethal concentration, and there is some comfort to be derived from the following quotation taken from a short but comprehensive review that Dr Southgate (1965) gave to the Salmon and Trout Association: 'An unexpected finding is that the greater part of the toxicity in the rivers so far examined is due to the presence of a very few substances. Phenol and ammonia are sometimes important but often the substances of overriding importance are copper, zinc, and cyanide. . . .'

Little was known about the fauna of British rivers until a governmental team started work between the wars. Their studies both of pure rivers and of rivers at different stages of pollution have not been equalled for thoroughness since, and remain classic. After the war Dr H. B. N. Hynes, travelling far as a consultant, took the opportunity to study stretches of many rivers. By then keys to several aquatic groups, ill-known previously, were available – Hynes was responsible for some himself – and therefore it is on Hynes' book that the following account is based. We start, however, with some observations on the Skerne, a tributary of the Tees, which was included in the survey of that river by Butcher, Longwell and Pentelow (1937). This, with the records from the Tees estuary, speak more eloquently of what pollution can do to a fishery than any general account based on many sources. Figures 24 and 25 are also based on the earlier work.

'The history of the fisheries of the Skerne is a striking example of the effect of pollution. In the early years of this

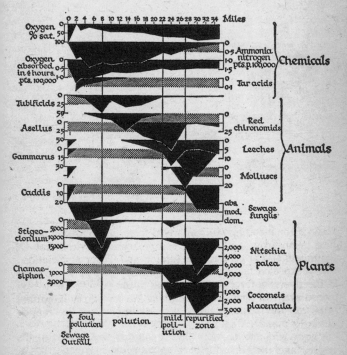

Fig. 24 Pollution in the River Trent

century it appears that this river contained large numbers of trout and coarse fish, and that it was improving, for in 1901 the Report of the Fishery Board records that for the first time in living memory sea-trout ascended the river and spawned at

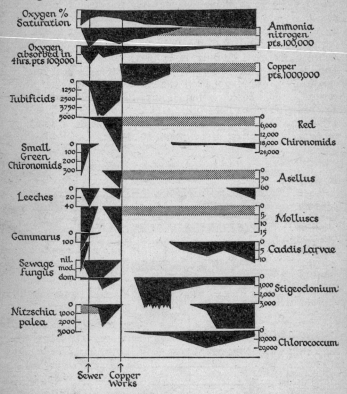

Fig.25 Pollution in the River Churnet

Darlington. They continued to do so until 1903 but there is no record since of their visiting this river. Almost annually the Board records some pollution which killed fish in the Skerne, but in 1906 and 1907 it was still considered worthwhile to stock the river with trout. After that it appears that pollution increased and the Fishery Board gave up the unequal struggle. Now the capture of a trout is sufficiently rare to be recorded in

the local newspapers. There is no doubt that in the absence of
undue pollution the Skerne would support a large stock of
fish.' (Butcher, Longwell, and Pentelow, 1937).

Though the freshwater reaches of the Tees are not seriously
affected, the estuary is in a very different state, and this has a
marked effect on the salmon which must pass through it.
Alexander, Southgate and Bassindale (1935) kept a watch on
the smolt migration of 1930 and picked up 2,246 dead bodies
stranded on the bank. It was impossible to estimate how many
were washed out to sea unrecovered or overlooked for some
other reason, or how many fell a prey to the gulls which could
be seen taking a heavy toll of the smolts, presumably those in a
moribund condition or at least suffering ill-effects to some
degree. The figure 2,246 obviously represents but a small fraction of the total casualties.

Table 10. Average number of fish caught at the mouth of the
Tees in nets

Period	No. of salmon caught	No. of sea-trout caught
1904–08	6,411	5,229
1909–13	8,067	5,816
1914–18	5,122	1,825
1919–23	5,730	4,423
1924–28	2,850	1,274
1929–33	1,826	816

The estuary receives the sewage from a population of some
280,000 persons in seven miles, and the oxygen may fall to 9%
of saturation. But during the period of smolt migration the
lowest concentration recorded was 50% of saturation, and it
was found that this was not low enough to be lethal. And
apparently the tar acids, which were present generally in a
concentration of between 0.1 to 0·3, rising occasionally to 0·5
parts per million, were not causing the death of the smolts
either. All the evidence pointed to the cyanides, of which there
were about 0.2 parts per million parts of water. Experiments
with invertebrates showed that some species were killed by this
concentration and others were not. The fauna of the estuary
was not greatly different from that of unpolluted but otherwise
similar estuaries elsewhere, though a number of marine species
did not penetrate as far into the Tees as into the unpolluted

estuaries (Fig. 19). In spite of the hazards provided by sewage, tar acids, and cyanides, some smolts were evidently, at the time of the survey, making their way to the sea, and some salmon were returning and penetrating to the spawning grounds. But the number has been decreasing since the turn of the century, and the tale is clearly told by the figures in Table 10. This shows the decline in the number of fish taken by nets at the mouth of the river; the numbers caught by angling had shown a similar but even greater reduction.

Turing (1947) continues the story: 'In 1937 the catch was 23; today it is nothing.'

Hynes (1960) writes that, when organic pollution is sufficiently intense to remove all the oxygen, there are few organisms except bacteria, among them the forms which produce long mucilaginous sheaths that cause the familiar unsightly whitish masses of so-called 'sewage fungus'. Two relatively large animals can live in these conditions because they make contact with the surface and take in air. One is the rat-tailed maggot, *Tubipora (Eristalis)*, often seen in, for example, shallow pools near a dung-heap or the outflow from an overloaded septic tank, places where the water is so turbid that nothing can be seen of the larvae except the tip of their telescopic tails rising from the murk and piercing the surface. Larvae of the other, the moth-fly, *Psychoda*, normally live in thin films of water, and trickling filters provide them with ideal conditions. Obviously neither of these animals can maintain itself in streams or rivers.

A gradual rise in concentration of oxygen in a sewage-laden river is accompanied by an increase in the variety of the larger invertebrates. First there is little but a population of worms, then come red chironomids, and, at a later stage, large numbers of the water hog-louse or water-slater, *Asellus*. Lower down there is an abundance of molluscs, leeches, and *Sialis*, and, as these become scarcer with further disappearance of the nutrients to which the sewage has by now been reduced, Ephemeroptera such as *Baetis*, many species of Trichoptera and Plecoptera increase in numbers to a point where the fauna is as it would have been without pollution.

Hynes also describes changes brought about by small amounts of sewage – enrichment rather than pollution – but his findings do not differ sufficiently from those noticed in Chapter 7 to justify devoting space to them.

Where the bottom is stony on account of a swift current,

sewage may lead to the appearance of sewage fungus and other bacteria but not to serious lack of oxygen. Under these circumstances *Gammarus* may persist, among a population of worms and chironomids, but it does not reproduce as vigorously as in clean water. On the other hand, most other species typical of a stony substratum disappear. Hynes suggests that they cannot cling to the stones covered with bacteria and he also notes that they become covered, and presumably incommoded, by bacteria. It must be supposed that animals which normally inhabit this zone are able to prevent this by means of an antibiotic secretion or in some other way.

Intense pollution may cause the disappearance of all algae but they soon reappear and reach maximum abundance in the same zone as *Asellus*. Common species on glass slides are *Stigeoclonium tenue, Nitzschia palea* and *Gomphonema parvulum*. These decrease in abundance further downstream and their place is taken by a wide variety of species dominated by *Cocconeis*, *Ulvella* and *Chamaesiphon*, algae found above the source of pollution. A greater variety of species is found on natural substrata and less is known about them; the provision of glass slides is a useful means of gathering information about pollution if not about the natural flora. A larger alga, *Cladophora glomerata*, often forms dense blankets in the regions where decomposition has produced a rich supply of nutrients. It is an undesirable growth for several reasons: if great masses are carried away by floods, as they often are, they may become entangled in and cause the uprooting of higher plants, they may de-oxygenate the water when they decompose, or they may strand, decay and stink. The beds are so dense that the respiration at night may use up enough oxygen to affect other organisms. They may also trap sufficient silt, if they endure long enough, to provide a substratum for higher plants.

Cladophora is a widely distributed species that flourishes in waters with a copious supply of nutrients. It grows thickly enough to become a pest where there is considerable enrichment, either in a river that has purified itself, as described above, or in one that receives the effluent from a large and efficient sewage plant. By converting the simple substances into organic matter once more and later decomposing, *Cladophora* causes a secondary pollution. Enrichment has become commoner in the last few years with a change in farming practice. Less dung is now spread on the land and more artificial fertilizer because, with wages at their present rates, the latter is cheaper. The dung, therefore, tends to fertilize the rivers rather than the land and a certain amount of artificial fertilizer is always leached by rain and conveyed to the nearest river.

Sewage provides an abundant supply of food and therefore those animals that can tolerate the accompanying low oxygen concentration are generally present in very large numbers. As only a few possess this ability, the fauna of intense pollution is characteristic. When conditions permit a more varied fauna the uniformity is less. There is keen competition among the species and heavy predation, and therefore a small change that favours one species to the detriment of another may produce a big shift in the composition of the community.

Where, it may be asked, did some of these animals live before man poured large quantities of sewage into streams and rivers? As Hynes points out, pollution by organic matter is not an unnatural phenomenon in the way that pollution by some products of industry is. Accumulations of dead leaves often decay fast enough to use up all the oxygen, and may even produce the foul conditions in which the rat-tailed maggot thrives. *Asellus* abounds particularly in overgrown ditches when there is much dead vegetation but sufficient flow to bring in the substances that facilitate decomposition and enough oxygen to satisfy its modest requirements. The snail-leech-*Sialis* fauna, is typical of waters that drain from soils rich in nutrients.

Figure 25 shows the fauna and flora of the River Churnet and the lower Dove. Over a stretch of ten miles not one single animal was found; plants were not affected so severely as this but in the first five miles they were very scarce. This was apparently due to a copper works, and analysis showed 1–2 parts per million of copper in the water.

The effect of lead-mining on the fauna of streams has also been studied carefully. The industry is a very old one in parts of Wales but, by the beginning of the century, most of the mines had fallen into disuse as they could not compete with bigger undertakings elsewhere. Some were reopened during the 1914–18 war but closed down again soon after it was over.

The rock is crushed and ground and then elutriated so that the lighter particles of stone are carried away and the heavier particles of galena or lead sulphide remain. Originally the washings were carried away by the nearest stream and, as far as can be made out from the records, this caused no trouble until machinery was introduced to do the crushing and grinding. Complaints then arose that the washings, which inevitably contain some metal, damaged pasture when they were deposited on the land by floods, and even caused the death of stock. To counter this the practice was adopted of sedimenting the washings before discharging the washing-water. This stilled the complaints but it gave no protection to the aquatic fauna, as the washing-water contains lead in solution. Furthermore the heaps of debris which accumulate from the sediment in the washing-tanks may yield enough lead, when rain-water seeps through them, to poison the stream for years after the mining has come to an end.

Laurie and Jones (1938) describe a survey of the lower reaches of the River Rheidol, in Cardiganshire, which drains an area in which there are no less than 43 derelict lead-mine workings. A few of these were opened up again during the 1914–18 war, and when Dr Kathleen Carpenter surveyed the stream in 1921–22 she was able to find only 14 species of animals. The composition of the fauna was interesting. All the animals belonged to the phylum Arthropoda and most of them were insects; among the insects the absence of caddis-flies (Trichoptera) was noteworthy. Analysis showed that there were 0·2 to 0·5 parts of lead per million parts of water.

By the following year there had been considerable recovery and Carpenter recorded 29 species, including Trichoptera and a flatworm, *Polycelis nigra*. The concentration of lead was less than 0.1 part per million, though it rose above this figure in floods.

Laurie and Jones made their survey in 1931 and 1932 and recorded 103 species, including molluscs and fish, which had been notable absentees in the earlier surveys. They believed this to be the normal fauna of a stream of the type of the Rheidol and accordingly concluded that recovery was complete. The concentration of lead was of the order of 0.02 parts per million rising to 0.1 part per million after heavy rain.

The authors point out that their figure of 103, which does not include the very small animals such as the Protozoa, is probably not completely accurate, since they were able to place many of the insect nymphs and larvae in the genus only and, in some genera, there may have been more than one species. They also adumbrate the possibility that the beds of these lead-polluted streams may contain silt and coarser particles, which are derived from the mine refuse dumps, and which would not be present in a stream-bed derived wholly from the process of natural erosion. At present, however, faunistic knowledge is not sufficiently detailed to make possible an assessment of what effect this factor may have on the fauna. J. R. E. Jones (1958), who incidentally has made an extensive study of the effect of poisons on fish (Jones 1964), suggests that the toxic effects that the earlier workers attributed to lead may have been due to zinc. He also puts forward the idea that abrasion by grit that would not have been there but for the mining and crushing operations removes algae which are the food of Trichoptera and some other scarce groups.

Carpenter (1926) recognizes four stages in the effect on life in a stream of pollution by a lead mine:

1. No life at all.
2. A sparse fauna composed almost entirely of insects, among which Trichoptera are notable absentees; a sparse flora with such algae as *Lemanea* and *Batrachospermum* and a few mosses and liverworts.
3. A richer fauna, including Oligochaetes (worms), Turbellaria (flatworms), and Trichoptera; flora including Chlorophyceae.
4. Fauna completed by the appearance of molluscs and fishes; flora by the appearance of higher plants.

These four stages may be seen in one river with increasing distance from the source of pollution, or at one point in a river with the passage of time.

The study of the effect of pollution leads on to the important matter of assessing pollution. A bacterial test is often carried out, and the purity of the water judged by the number of bacteria whose real home is the human intestine. This is a simple and reliable method of detecting pollution from sewage, and it has the advantage that it also gives a measure of the degree of contamination. But the drawback to it is that a negative result tells nothing. There may easily be intermittent pollution which is sufficient to harm the fauna but which will not be detected by the bacteriologist if his sample is not taken at the right time; and perhaps the offender is releasing his effluent only after dark or at some other time when he has reason to suppose that no-one with a sampling-bottle is prowling near the river. Moreover it is practicable to mix sewage with some bactericide, usually chlorine, so that decomposition is delayed for a period. This practice is not uncommon when, in the interval, the sewage is carried out to sea, but it is also available to the unscrupulous person seeking to avoid a summons for pollution. Further, the bacteriologist will not detect pollution from sources other than sewage; though possibly, since copper is a good bactericide, the extremely pure condition of the Churnet (see Fig. 25) below the copper works might have led an astute observer to suspect something unusual.

The chemist can cover a wider field. He tests for pollution with organic matter in various ways. First, and most important, is the amount of oxygen in solution in the water; but it is not always possible to test for this, because certain reagents must be added to the sample immediately after it has been collected. The actual amount of organic matter may be determined. In a sample of water left to stand, organic matter will be decomposed and the process will use up oxygen. The chemist measures the amount of oxygen which disappears in a given time at a given temperature, the standard generally accepted now being five days at 20° C., and assumes that it is proportional to the amount of organic matter. He calls his answer the B.O.D., which stands for biological or biochemical oxygen demand. The amount of oxygen taken from potassium permanganate in a given time at a given temperature is a chemical reaction giving some indication of the amount of organic matter present; it is found to be a useful test in practice though little is known about what exactly does take place. The amount of ammonia is also an indication of the degree of pollution. How the concentration of oxygen rises, and the B.O.D. and ammonia concentration fall as a polluted river gradually repurifies itself is shown in Figure 24.

The chemist can also test for other known toxic substances such as metallic salts, and a chemical analysis to examine for the presence of every substance known to have a harmful effect on freshwater life is well within the bounds of possibility; but it is a lengthy process. Moreover it can never be ruled out that some hitherto unknown substance is causing the trouble; and the chemist may be misled in the same way as the bacteriologist by a sample not taken at the proper time.

With increasing risk of pollution from something new and unknown, increasing use is being made of caged fish, much as miners once detected the presence of harmful gases by means of a caged canary.

If pollution is intermittent, variable in intensity, or cumulative in action, the best indicators of its effects are the animals and plants. Moreover surveys of the communities at regular intervals afford a useful means of finding out whether the condition of a river is improving, deteriorating, or remaining unchanged. The drawback to this method is that it requires somebody who, familiar with the fauna and flora in all parts of clean rivers, can make a reliable assessment of what would be found at a certain place if there were no pollution, and such people are not numerous.

This chapter is being revised at a time when pollution is news. The President of the United States and the Prime Minister of Britain have both recently denounced it and promised legislation against it. Their statements have achieved the doubtless desired end of considerable newspaper publicity; *The Observer* of February 2nd, 1970 contained two articles and a leader on the subject, and *The Times* followed next morning with its first leader. The press labours under the disadvantage that the need to publish news while it is new does not leave time for the precise checking of facts that a scientist demands, and it is probably inevitable that the public is served with a certain amount of misinformation in articles on topics of concern to all citizens. For example, *The Times* leader contains the following sentence: 'The Thames, so polluted as to be fishless 20 years ago, has now some 40 species of coarse fish.' If any angler who had been taking fish from the Thames for the last 50 years wrote to refute this, and to point out that there are not 40 species of coarse fish in Britain, the editor ignored the letter. How different it would have been if he had stated, or permitted a reporter to state, as the editor of the *Westmorland Gazette* recently did, that Beatrix Potter wrote *Winnie the Pooh*. All involved were gravely concerned at this literary mistake, a number of letters were published and an apology was offered. The purpose of this digression is to illustrate our contention that a man who would be deeply offended by any suggestion that he was not well educated can admit without a blush that he has no knowledge of biology or, venturing into that field, make errors of a kind that would cover him with shame if he made them in any other. As so many of the problems which beset human civilization are biological in origin, we believe that this attitude should be changed.

Had the sentence above been altered by the removal of the word 'coarse' and the insertion of 'estuary' after Thames, it would have been accurate. A vigorous attempt to reduce the pollution of the Thames in London has been made recently and, though the load is still heavy, the condition is better than it was. Southgate (1969) has described the operation. Wheeler (1969) has given an account of the effect of it on fish, together with a useful history of pollution and fisheries in the Thames. Both Wheeler himself and F. T. K. Pentelow made enquiries in 1957 among anglers, waterside workers, naturalists, and the inspectorate of the Ministry of Agriculture, Fisheries, and Food. These indicated that there were no fish below Richmond (26 km. above London Bridge) and above Gravesend (42 km. below it). It is hard to prove a negative in the field. If fish are not caught, there may be none to catch, but it is always possible that the wrong methods are being used. However, there was an undoubted change as the condition of the river improved, for in 1964 and 1965 fish were found on the screens over the intakes to power station condensers in the stretch between Richmond and Gravesend. In a year starting from September, 1967 Wheeler obtained fish trapped in this way from five power stations, the highest 9·6 km. above, the lowest 35·4 km. below London Bridge. Eleven species of coarse fish were taken in the upper parts of the zone, and twenty-three marine species in the lower. There were also seven estuarine species.

Even when *The Times* statement has been corrected, it presents a picture that is too rosy. On the other hand many newspaper articles have painted a picture that is too black; indeed some give the impression that pollution is getting worse sufficiently fast to menace our civilization. Dr B. A. Southgate was, until his recent retirement, Director of the Water Pollution Research Laboratory, and his opinion is likely to be the most informed. He writes (1969): 'It is likely that the proportion of fishing waters will increase rather than diminish in the next 20 years or so,' and again, to conclude the book: 'The last general survey of the condition of British rivers was that, previously mentioned, made in 1958; if another were made now the author would predict that their state would be found to be better than it was then.'

It is known how to render harmless any substance likely to be discharged into a river, and hitherto it has rarely taken long to discover how to treat any new source of pollution. Pure rivers are, therefore, a social and political issue. How much are we prepared to pay for them? Any effluent can be rendered harmless, but the cost of doing this may add a significant amount to the final price of the article being manufactured. Would it price it out of the market? Are we prepared to lose business to some country in which there is less concern about natural amenities? These are the sort of questions which must be answered before the issue is finally judged.

The Rivers Pollution Prevention Act of 1876 applies to all nontidal waters and to any tidal waters which have been declared a stream by the Minister of Health. It makes it an offence to discharge into a stream solid matter so as to interfere with the flow of the stream or to pollute it with sewage or factory waste. But there is an escape clause to the effect that no offence is committed if the best practicable and reasonably available means are being used to render the discharge harmless. Any measures to render an effluent harmless will cost money, and it was successfully argued that an expensive outlay for this purpose did not come within the meaning of 'reasonably available'. This was a cogent argument in the seventies – the heyday of industrial expansion and free competition – and today the whole populace is conscious as never before of the vital part which industrial production plays in the life of the nation and of the importance of low costs. None the less it is difficult to resist the conclusion that the Act of 1876 weighted the scales unduly against those interested in keeping rivers pure; and that many concerns used the escape clause to avoid installing purification plant which they could well have afforded.

The enforcement of the Act was placed in the hands of the local authorities, most of whom made little use of the new power bestowed on them. One may surmise that they were unwilling to burden the rates with the expense of installing a sewage-purification system in response to what they perhaps stigmatized as a new-fangled whim. It is interesting to note (Frankland and Morton, 1874) that in 1874 the Commissioners, appointed in 1868 to inquire into the best means of preventing the pollution of rivers, came to the following conclusion about cholera and typhoid, now of course known to be caused by bacteria: 'The existence of specific poisons capable of producing cholera and typhoid fever is attested by evidence so abundant and strong as to be practically irresistible. These poisons are contained in the discharges from the bowels of persons suffering from these diseases.' The Commissioners were enlightened men, and their advanced theories probably did not impress local councillors, who possibly were not. Furthermore, if the pollution of a river led to difficulties over water supply, that was a problem for the next local authority down the stream and no concern of the one responsible for the trouble.

The duty of enforcement was transferred to County Councils when these were formed under the Local Government Act of 1888, and this did produce some improvement, as the jurisdiction of a County Council extended over long stretches and sometimes over the whole of a river. The Local Government Act also made provision for the setting up of joint committees to administer the Rivers Pollution Prevention Act, and where these were formed they did achieve a measure of success.

The general position, however, remained highly unsatisfactory, and in 1898 the Government appointed a Royal Commission on Sewage Disposal. This set about its duties in a thorough manner and initiated original research when it could not obtain from any other source information which it considered important. The final report was issued in 1915. The commission recommended certain standards, for example the desirable dilution for sewage described above, and gave an account of methods by which different kinds of effluent could be rendered innocuous. The middle of a war was, however, an unfortunate time for the issue of a report of this sort.

The Salmon and Freshwater Fisheries Act of 1923 made it an offence to discharge into tidal or non-tidal waters any substance liable to injure fish or affect adversely their spawning-grounds or food.

Various persons and bodies interested in river purity campaigned vigorously between the wars. In 1921 the Salmon and Trout Association convened a meeting at Fishmongers Hall and the outcome was a deputation to the Minister of Agriculture and Fisheries, who appointed a Standing Committee on River Pollution. This committee, which was allocated no funds, made surveys and could call on the technical experts of the Ministry of Agriculture and Fisheries, which, at about this time, equipped a laboratory and appointed some scientific staff. The committee aimed to prevent pollution wherever possible, by amicable arrangement when this could be arranged and by legal action through the Fishery Boards when it could not. They were concerned with preventing fresh pollution rather than with an assault on the entrenched strongholds.

In 1926 the Salmon and Trout Association and the British Waterworks Association drafted a memorandum and sent it to Mr Stanley Baldwin. In the following year, with the Fishmonger's Company, they organized a deputation on which almost every one of the many diverse interests in water was represented. The result of this deputation was the setting up of a Central Advisory Committee and a Water Pollution Research Board, the latter to continue the work of the Royal Commission which had dissoved in 1915. This was a particularly necessary step, not only to investigate ways and means of dealing with effluents of various kinds and to seek continually ways and means of improving existing methods, but also to keep on terms with advances in industry and possible new sources of pollution. Some idea of the scope of the Water Pollution Research Board may be gained from the fact that it issues monthly a summary of current literature, each number summarizing between 100 and 200 papers.

The important changes of subsequent legislation have been to increase the area for which a single body is responsible and to bring control of drainage, fisheries, and public health under one head. The most recent act in this series is the Water Resources Act of 1963. River Authorities were established and to the duties already mentioned was added that of conserving water. The same act established the Water Resources Board, to keep under constant review the water resources and water requirements of the whole country. River Authorities are charged with the laying down of standards for each effluent discharged into any water under their control. This they have found far from easy. It took the Royal Commission of 1898 seventeen years to lay down standards for sewage effluents. Modern techniques and knowledge should make it possible to complete a similar investigation more quickly today, but obviously an answer can be expected only after years, not after months. There are many substances to test and the possibilities exist either that they exert a cumulative effect over a long time, or that they combine with some other substance to give a poisonous product. Quick answers from River Authorities today will, therefore, obviously mean recriminations tomorrow. To reduce pollution they have tended to approach the parties in a friendly and collaborative spirit rather than to invoke the might of the law.

At present sewage works, many of which receive and treat industrial effluents are the responsibilities of Local Authorities. It has been suggested that they too should be administered by the River Authority, in order that one body should have control over sewage purification and prevention of pollution.

Nearly a century of legislation has failed to save many of

our rivers from pollution, but if Statute Law has not been provided with teeth, Common Law possesses them. At common law a riparian owner has a right to receive water unaffected by the activities of persons higher upstream (there are a few small exceptions to this) but the expenses of going to law are so great that few private citizens like to contemplate this step. If, however, costs in the event of defeat could be shared among many anglers, the burden on each would be small, and a victory by one individual would be in the interests of all. This, together with the provision of technical assistance and legal advice, was the basis on which the Angler's Cooperative Association (A.C.A.) was founded. It was soon in action. There were anxious moments. Powerful defendants took cases to the High Court and the Court of Appeal and the loss of a big case could have proved disastrous. But the A.C.A. went from success to success and many miles of river are now free of pollution as a result. The Association issues a handbook which contains a clear and simple account of the various acts that Parliament has passed and of the legal situation today.

15. Pure Water

The subject of water supply can be approached in two ways by the naturalist: how do the inhabitants of fresh water affect the engineer; and how does the engineer affect the inhabitants of fresh water?

The congregation of population in great conurbations, coupled with ever advancing ideas about hygiene and cleanliness, has made water supply one of the major features of modern civilization. It is easy to imagine that, if in a thousand years' time our civilization is represented by nothing but ruins, great concrete reservoirs, or vast masonry dams blocking valleys, or even the conduit tunnels bored through the hills, will be among the more imposing of the remains.

Water supply is not only a matter of conveying water in the required amount to the required place at the required time. The water must also be purified, for it can be a dangerous substance to drink. The traveller who, in tropical lands, is not careful about where his water comes from, runs the risk of contracting a number of serious diseases. Notoriously rampant in Egypt is a small worm of the fluke type known as *Bilharzia*, a name which has been simplified to 'Bill Harris' by the British Army, sometimes to be relatinized into 'Billharrisia' by persons who wish to parade a show of knowledge. *Bilharzia* lives in the abdominal veins of man and there it lays its eggs. These work their way through the tissues to the bladder or the rectum, according to the species, and are passed to the exterior. There is no further development unless they reach fresh water, but the chances of this happening are high in a country where the local populace has no knowledge of hygiene. A larval stage comes out of the egg and swims off in search of a certain kind of snail. If the search is successful, it enters the snail, penetrates the liver, and there reproduces asexually, giving rise to a great number of young. A small swimming stage emerges from the snail and it reinvades the human host, penetrating any part of the body if it is not taken in at the mouth. Another disease, also interesting to the zoologist, is caused by the guinea-worm which, when adult, lies in the tissues in a tube which communicates with the surface for egg-laying through a small ulcer. The creature, which may reach a length of 40 inches (100 cm.), is commonly removed with the aid of a matchstick on which it is rolled, a little bit being extracted each day. Its other host is a small water-flea, and the life-cycle is completed only when a human swallows an infected water-flea while drinking or bathing.

Then there are the diseases caused by bacteria: cholera, fevers of the typhoid group (which are not to be mistaken for typhus, a disease carried by lice), and the complex of complaints which goes by the name of dysentery. Recently some evidence has been brought forward that certain disease-causing viruses may pass from man to man in water, and this possibility is under investigation.

In Britain the only diseases known definitely to be waterborne are due to bacteria and there is nothing higher in the biological scale which will harm human beings. Of the three mentioned, cholera is now unknown in these latitudes and has been so for more than seventy years, though as recently as 1874 it carried off 20,000 people in an epidemic. Danger from typhoid and dysentery is never completely absent, though both are now rare. Even a few cases of typhoid provide headline news today, though a century ago epidemics with a death-roll of thousands were only just beginning to excite comment. Water, therefore, is safe to drink in Britain if it is free of disease-bearing organisms. But the waterworks engineer has to go farther than this, for what is good for the consumer and what the consumer wants are not identical. The layman seldom worries unduly about bacteria, for what the eye cannot see the heart does not grieve about, but he does demand that his water shall be colourless, tasteless, and free of any living thing which can be seen. Nothing, we understand, leads more quickly to an angry letter or to an enraged caller at the waterworks office than the emergence of something alive from the domestic tap. The waterworks engineer must, therefore, solve the problem of rendering water free of taste, colour, and all visible matter, dead or alive. The consumer is always right.

Water may be taken from three main sources – underground supplies, lakes or reservoirs, and rivers; and there is a different set of problems associated with each one. Underground supplies are usually free of bacteria and all other forms of life, for percolation through soil is an efficient method of filtration. But it is a mistake to suppose that underground water, even though it comes from quite deep down, is necessarily pure. Sometimes there are cracks and fissures through which water polluted at the surface may run straight down into some subterranean reservoir without any percolation through soil.

Modern methods of testing the purity of water depend on ascertaining the number of *Escherichia* (*Bacterium*) *coli* per litre. *E. coli* is an inhabitant of the human intestinal tract and can live for quite a long time in water. Although itself harmless, its presence, therefore, indicates probable contamination with human excreta and implies a potential risk that bacteria which cause disease may be in the water too. A standard assessment of purity is:

> Under 20 *E. coli* per litre—good
> 20–100 „ „ „ „ fair
> Over 100 „ „ „ „ bad

Complete absence is not demanded, as a certain number are derived from animals other than man.

The engineer who can tap a subterranean supply of water is likely to be faced with the problems of hardness, oxygen deficiency, and richness in nutrient salts. Hardness is a purely chemical problem. The chief effect of oxygen deficiency is to bring into solution any iron present in the soils through which the water has come. Certain bacteria oxidize soluble ferrous compounds back to insoluble ferric iron, and, if there is much organic matter present, they may be extremely abundant. When water rich in organic matter and ferrous iron is led through pipes, a gradual blocking up of the pipes by the deposition of ferric iron will be caused by the activities of the iron bacteria. Furthermore, the water which reaches the consumer will be stained a rusty colour, and a dripping tap will leave unsightly stains on bath or wash-basin.

If underground water is rich in plant nutrients it should not be brought into an open reservoir or there will be a crop of phytoplankton to be coped with.

Many large towns within reach of Scotland, Wales, the Lake District, or the Pennine Chain derive their water from some reservoir in the mountains; either utilizing an existing lake, which is often increased in size by a dam, or by blocking a valley and creating an entirely artificial reservoir. Some of the lakes are unproductive and yield almost pure water; the dissolved substances in Thirlmere, Manchester's supply, are similar to those in Ennerdale Water given in Table 1, on p. 33, and the water could safely be supplied straight to the consumer without any treatment at all. Actually, to guard against remote possibilities, the water is chlorinated. Other lakes are more productive and accordingly less ideal as sources of water, for the greater amount of life means that the water must be filtered as well as chlorinated; the richer the lake the greater the work and therefore the higher the costs at the filtration plant.

Apart from the high initial cost of a long pipeline, a reservoir in barren upland country is ideal in the eyes of a city corporation; danger from disease is remote, the expenses of filtration are low, and gravity does the work of conveying the water to where it is required. It is not ideal in the eyes of another section of the community, and we may relate the history of Thirlmere to illustrate the resulting conflict, adding that this lake is typical of England only; similar treatment of Welsh lakes and export of the water for use across the border has inflamed nationalist feelings and provoked sabotage. Manchester Corporation acquired Thirlmere and much of its drainage area towards the end of the last century and enlarged it by building a dam. They fenced it and ordered the public to keep out. They planted conifers on the surrounding fellsides to prevent the erosion which enriches a lake. Both measures annoyed lovers of the Lake District, and annoyance has swelled considerably in recent years as demands for amenity have grown in step with the rise in the general standard of living. Many more people have the leisure and the money to fish, to bathe, and to mess about in boats, and in this small and crowded island any restriction, particularly in a National Park, as many of the areas suitable for upland reservoirs now are, is resented. Amenity is, of course, a difficult item to show in a balance sheet to be scrutinized by heavily taxed ratepayers; the walker who asks no more than to be left alone to wander where he will, enjoying the solitude if he can find it, and the scenery, is certainly not going to pay for what he holds to be his right. There are, however, others from whom money can be extracted, and official attitude is changing from that of an old-style landowner to one which seeks to offset the extra costs of purifying the water from a well-used reservoir by providing facilities and charging for them. For example, Grafham Water, near Huntingdon, a reservoir opened in the mid-sixties, was planned to provide facilities for enjoyment of the pastoral scene, for fishing, for sailing, and for the protection of nature in a reserve. Those who enjoy the last three facilities pay for

them. It is impossible to please everyone. There are now some
who have a good word to say for Manchester Corporation, and
thank them for one lake where points of easy access are not
scarred by trampling, where plastic containers are not common
objects of the lake shore, and where the valley walls do not
reverberate with the noise of outboard motors.

From the amenity point of view, the best place to take water
is just above the estuary, where it is about to lose itself in the
sea and nobody can derive any further enjoyment from it.
Operating costs, however, are bound to be higher because the
water must be purified and then pumped to where it is required.
The problems have long been familiar to water suppliers in the
south and east of England, who have had to take water from
rivers that have been used for many purposes, including the
discharge of sewage, higher upstream. London derives much
of its water from the Thames and other rivers.

Large reservoirs are an essential feature of a water-supply
system derived from rivers. They store a reserve which can be
drawn upon at times of greatest demand and they also have an
important function as sedimentation tanks for the mineral
matter which every river carries in suspension. Some intestinal
bacteria can survive for a period in fresh water but they cannot
reproduce to any great extent, and consequently storage in a
reservoir greatly reduces their numbers. But the still, clear conditions, with a big surface area, are ideal for the growth of
phytoplankton. Phytoplankton leads to zooplankton and to
larger animals as well, though, in a reservoir with sides running down sheer to a depth of several feet, conditions are not
very good for these last, provided that rooted vegetation is not
allowed to grow.

These living organisms must be strained off before the water
is fit for domestic supply, and this is generally done by passing
it through a bed of sand. Pressure may be applied (either by
means of a big head of water or a pump), or it may not. These
two methods, rapid and slow sand filtration respectively, are
almost always augmented by some other process.

Chemical precipitation is often associated with rapid sand
filtration. The water is first treated with alum or some other
substance which, on contact with water, gives rise to a colloidal
hydroxide of iron or aluminium. A jelly-like net is formed in
the surface layers and sinks to the bottom, taking with it all
particulate matter, living and dead. The water is then passed
through the sand bed.

A slow sand filter is commonly used in conjunction with biological filtration and purification. The process takes place in a rectangular basin not unlike a swimming-bath to look at. Gravel is placed on the bottom and above it a graded series of sands, two to three feet deep, the finest being at the surface. Water is run into the basin and allowed to stand for a few days. The result is a rich living growth. Various algae, particularly diatoms, bacteria, and protozoa, form a carpet on the surface of the sand, and larger animals, particularly chironomid larvae, may inhabit this carpet in great numbers. Underneath it bacteria grow on the sand grains, which they surround with a gelatinous covering. This complex of living organisms is the effective part of the filter. It exerts a mechanical effect and planktonic algae and animals are strained out on its surface. It also oxygenates the water and this is an advantage for several reasons, not least because water with plenty of oxygen dissolved in it has a sparkle which is pleasing to the palate of the connoisseur. Much organic matter in the water is oxidized and broken down into simple compounds, and these, with inorganic substances already present, are removed by the algae. It seems likely that some part of the system, possibly the gelatinous envelope round the bacteria in the sand, is absorbent, for traces of substances such as copper which have been used to kill algae in the reservoir are removed by a slow sand filter.

After filtration the water is chlorinated, for there is always a chance that a few bacteria may pass the filter; and then it is ready for supply to the consumer.

A filter-bed clogs up sooner or later, sooner if the water contains many diatoms, for the skeleton of these plants is relatively indestructible. Some species appear to be a good deal worse than others. The surface layers of sand are then removed and cleaned and the whole process starts afresh. Pearsall, Gardiner, and Greenshields (1946), to whom reference may be made for a fuller account of biology in relation to waterworks, give some figures which bring home vividly the problem which the waterworks engineer has to solve. A reservoir receiving water from the Thames and holding about 7,000 million gallons might, at certain times, contain 110 tons dry weight of the alga *Fragilaria crotonensis*; the wet weight would be of the order of 1,000 tons. It is interesting as a sideline to make a comparison here with Windermere, in which a similar volume would contain less than 3 tons, dry weight, of

Asterionella at the height of the season. The engineer supplying water from this reservoir at a normal rate would have to remove about 10 tons wet weight of *Fragilaria crotonensis* every day.

The engineer can take certain measures to avoid this load and, at this point, theoretical and applied limnology meet. A reservoir is a lake, and the scene of the various events outlined in Chapter 2. It stratifies in summer into a warm upper (epilimnion) and a cold lower layer (hypolimnion). If it is productive, the organisms produced in the upper layer fall to the lower layer and decompose there, using up all the oxygen. Under these circumstances iron goes into solution and water rich in iron should not be passed through pipes because, when it is, iron bacteria precipitate a rust-like incrustation which comparatively rapidly blocks them. The hypolimnion will, however, become de-oxygenated only if it is small. The ideal reservoir should, therefore, be deep. From such a reservoir the engineer may safely draw his water from the region of the hypolimnion at a depth too great for the multiplication of algae. This ideal, however, is not easily attained, and the engineer has often to do the best he can with a rich and shallow reservoir. Under these conditions the water should come from above the hypolimnion, but as deep as possible, because it is near the surface that light conditions are optimal for algal growth. A complication is that the thermocline, the region between epilimnion and hypolimnion, is tilted by the wind. Water is so heavy that at the lee end of a lake the water level is scarcely higher than at the weather end when a gale is blowing, but the epilimnion may be driven downwind in a wedge-shaped mass, with the thermocline tilting, sometimes to such an extent that the cold hypolimnion is exposed at the weather end of the lake. When the wind drops, the epilimnion flows back, overshoots, and piles up at the opposite end of the lake. The thermocline tilts up and down exactly like a see-saw and this process continues though each tilt is slightly less steep. If the outlet from the reservoir is at one end, this seiche, to use the technical term, will bring water with the maximum concentration of algae to the pipe when the thermocline is tilted in that direction. As it rises the more suitable deeper water comes to outflow level, but as the tilt approaches the top of its travel the hypolimnion water without oxygen runs into the supply. If the reservoir is long, the seiche will tend to tilt from end to end and an outlet near the middle of one side, in other words opposite the point where the see-saw is supported, will receive water of constant quality. However, in many reservoirs the seiche may swing in any direction according to where the wind is blowing from, and a constant supply of best quality water can be drawn from them only through outlets situated at different levels and brought into use at the appropriate time. The Metropolitan Water Board, for instance, have installed an elaborate system of recording the various factors mentioned, in order to be able to draw off the most suitable water all the time. The population of an alga generally increases suddenly and then drops suddenly. If an undertaking has several reservoirs, it can sometimes leave one in which an alga is reproducing rapidly untapped until the outburst is over. If this is not possible, load on the filters may be saved by poisoning the alga as soon as examination has revealed that an outburst is starting. Copper sulphate is most commonly used and quite low concentrations are lethal to algae. Any copper left over is apparently removed by sand filtration, though the sulphate may lead to trouble later, for, if all the oxygen is used up, it may give rise to sulphides.

Algae are troublesome to the engineer not only because they block his filters; some impart a distinct taste to the water, usually an unpleasant one. Sometimes it is highly resistant to treatment but usually it can be removed fairly easily, by aeration, by the use of oxidizing agents such as ozone, or by passage through special charcoal screens.

Some of the earlier water undertakings conveyed water in pipes without treating it, which led to troubles vexatious to them but fascinating to the naturalist. The most detailed study was that made of the Hamburg supply system; this is briefly summarized by Dr A. B. Hastings (1937) in a British Museum pamphlet on *The Biology of Water Supply*. The frequent but never excessive flow makes the inside of water pipes a particularly suitable place for those animals which are sedentary and which obtain their food by filtering particles out of the water. Since such forms usually have a motile stage early in the life-history, their chances of attaining this environment are good. The Polyzoa, or moss-animalcules, live in branched colonies not unlike certain corals. Sponges which at certain times of the year produce resting spore-like bodies known as statoblasts, and *Dreissensia*, the zebra-mussel, which can attach itself to a hard substratum and is not a mud-dweller as the other bivalves are, were all common in pipes. *Dreissensia*, which has invaded fresh water from the sea quite recently, still has a planktonic larval stage. This is common to all bivalves which live in the sea but has been lost in every other freshwater species. Pipes are not suitable for animals that fly in the adult stage and therefore do not harbour the larvae of *Simulium* and various net-spinning Trichoptera that derive their food from fine particles carried in streams and rivers.

Once the walls of the pipes are well lined with Polyzoa, sponges, and *Dreissensia*, there will be an abundance of nooks and crannies in which other animals can find harbourage, and various Crustacea, particularly *Asellus*, the water hog-louse, may become established. It is a debris-feeder, and presumably finds a good supply of food lodged in the many holes and crevices which the overgrown surface of the pipe now presents. The basis of a food-chain is thus firmly established, and in the Hamburg water system even eels were abundant. The Hamburg water pipes supported a rich and varied fauna until sand filtration was introduced when it died off and disappeared.

Recently reports of animals in pipes have become more frequent, though whether this represents a real increase or a less reticent generation of engineers is hard to determine. *Asellus* appears to be the most widespread and abundant inhabitant and Houghton (1968), whose paper is the most recent that has come to our notice, states that the experience of himself and of others is that *Asellus* has established itself after a change from slow to rapid filtration. This, together with the records of numerous other species found occasionally, suggests that many animals are entering the systems and that *Asellus* is one of the very few that can thrive there. What peculiarity enables it to do so is not known. H. P. Moon examined the gut contents of some specimens and found that they had been feeding on iron bacteria. Houghton thinks that enough algae pass through the filters to feed the population. Possibly some aspect of its behaviour is responsible for the success of *Asellus* in this biotope. Pyrethrin is commonly used to destroy populations in pipes.

These are some of the many ways in which the inhabitants of fresh water impinge upon the engineer. How does the engineer affect the inhabitants of fresh water? On the whole his activities are beneficial. We cannot do better than quote two sentences from Fitter's (1945) *London's Natural History* to support this statement: 'The creation by the Metropolitan Water Board of an extensive series of large lakes in the valleys of the Thames and Lea has been one of the major events in the recent development of London's avifauna. In an average winter practically every duck, grebe, and diver that visits the British Isles regularly may be seen with tolerable certainty on one of the reservoirs within twenty miles of London.' Obviously this is also true for many invertebrate animals. The all-concrete storage reservoir has no counterpart in nature, and will therefore harbour only a portion of the freshwater fauna and flora, but the impounding reservoir formed by damming a valley may differ in no important respect from a natural lake in the same surroundings. Water supply therefore leads to a greater number of pieces of water for aquatic animals and plants to live in. But sometimes the amount of water drawn off from a reservoir leads to a great fluctuation in the water level; there will be a large and irregular rise and fall depending on the rainfall, and there may be a smaller diurnal rise and fall caused by a fluctuating demand during the twenty-four hours. The result is that wave-action may extend over a greater vertical range than in a natural lake, and the establishment of rooted vegetation in the deeper parts of the shallow-water region cannot take place. This has far-reaching repercussions, and the result of the fluctuation is an impoverishment of both fauna and flora.

16. Summary

An attempt has been made in these pages to compose a well-balanced unified picture; in this, the summary, we shall stress the gaps and empty spaces existing within the framework of the title *Life in Lakes and Rivers*.

The underlying theme is based on four interrogatives, *what? where? why?* and *how much?* Chapter 6 is devoted to answering the question *what* occurs in fresh water. An exact identification of the species concerned is required as a preliminary to most types of work, and at this early point the first gap in our knowledge becomes apparent. The greater part of the fauna in some places is composed of insect nymphs and larvae, but not every one of these can at present be identified with certainty. All, except perhaps those belonging to a few big groups, probably are distinct, but the differences between them still await the worker who will seek them out and describe them.

Excellent progress has been made with this work during the last twenty years and on present form another twenty years should see a reliable key available for every freshwater group. At the present moment (early 1970) the Freshwater Biological Association has issued twenty-four special publications, most of them keys, and there are three ready for the press. Macan (1959) has written a guide to freshwater invertebrates that takes those unfamiliar with our fauna down to a group and indicates what keys, if any, continue down to species.

The question *where* is answered by those who select some particular biotope for study. The first task is the production of a list of species, and this becomes much more valuable if the life histories of at least the common species are known and there is information about their abundance at different times of the life history. This kind of work has not been popular in the past, partly because it was difficult or impossible to name or get named so many of the specimens caught. It has been increasing in popularity in recent years and it is to be hoped that the recent creation of the Natural Environment Research Council will provide further encouragement. As we pointed out earlier, there is a great deal of talk about conservation but some very large gaps in our knowledge of what we are conserving. For example, there have been many recent studies of small stony streams, but in these pages we had to refer to work done in the thirties when we wished to discuss the fauna of rivers. Similarly the Lake District lakes have been studied, but for a contrast with productive calcareous conditions we had to take the Danish Lake Esrom. Smaller bodies of water still offer a wide field for exploration.

Investigation along the lines described in the preceding paragraph opens the way to an answer to the other two questions. The question *how much* ushers the investigator into the most fashionable circle at the present time. As was stressed in Chapter 9, attempts to draw up an accurate statement of gain and loss are still completed only with the aid of a few assumptions of which some are scarcely more than guesses. At a less ambitious level the freshwater biologist can provide useful information to the waterworks engineer, who is interested in reducing production, and to the angler, who seeks to increase it. The applied field of fish culture is of the greatest importance in tropical lands where protein is in short supply.

When we try to answer the question *why* – why does species A occur in locality B, but not in locality C, or why does lake D support so many more animals than lake E – we quickly pass far beyond the confines of water and of biology. Chemical problems present themselves first and immediately lead on to others. Circulation in a lake affects the distribution of nutrient substances, and therefore the study must branch out to include the sun and its heat, the wind, and the properties of the water. But the nutrient substances are derived mainly from the drainage area, and search for their origin leads out of the water into the realms of geology and climate. Bound up with geology are agricultural practice and industrial practice, both factors which may exert a major influence on fresh water. Professor Pearsall and Dr Pennington, whose interpretation of the bottom deposits of Windermere was described in Chapter 4, have gone further still and brought archaeology and history within the orbit of the freshwater naturalist.

The answer to many of these *why* questions is to be provided only by a well-equipped laboratory and a team of research workers with a wide knowledge of physics and chemistry, and a flair for designing and constructing elaborate apparatus. But the territory is not by any means closed to the amateur, and the complicated answer is not necessarily the correct one. The mosquito which deserted the rice-fields at a certain stage of development because she liked to fly over the water and drop her eggs from the air is a good illustration (p. 166).

It is probably true to say that there is not a single piece of water in the country unaffected by man. Mountain torrents and high tarns and pools might be put forward by some to dispute this statement, for few of them are directly influenced by man today. But, had not man cut down the trees around them long ago and prevented regeneration by running his sheep over the land, they would not be exactly as they are today. Therefore freshwater natural history becomes inextricably mixed with the social organization of the country, regrettably no doubt, because many, both fishermen and naturalists, look to fresh water to provide them with an escape from the conflicts and worries of everyday modern life. But the issue must be faced.

In these pages we have dealt with the claims of four interests in the lakes and rivers of this country, claims which coincide and conflict in a complex pattern. First there are the suppliers of water, seeking new sources of supply as the demand rises. Subterranean water is showing signs of having been exploited to capacity, and attention is now increasingly directed towards new reservoirs, utilization of existing lakes, and tapping of rivers. The daily rise and fall in a reservoir may have an unfavourable effect on fish food, and the compensation water compelled by law may not satisfy the fisherman whose river water is taken for domestic supply; on the other hand a new reservoir may provide a new fishing-ground. Fishermen and waterworks officials are therefore sometimes friends and sometimes foes. The local authority disposing of sewage and the industrialist disposing of trade wastes are popular with few; but we depend on the industrialist for our very bread today and the pros and cons of adding to his costs by demanding extensive purification schemes must be carefully weighed. The fisherman will alter a river on his own account if he can, but he will not come into serious conflict with anyone, apart from the polluter, except perhaps over the question of access. The naturalist, if he is not also a fisherman, is probably interested to leave a place as little altered as possible. Nobody's claim is paramount; a compromise is the only solution, and it behoves the fisherman, the waterworks official, the waste products disposer, and the naturalist each to acquaint himself with the claims and problems of the others. We conclude this book with an expression of hope that some have found it a useful aid to unprejudiced opinion about the future of our natural environment.

Bibliography

Twenty years ago we built up most of the chapters from information in original papers. Today most of them are covered by books. Macan (1970) has written an account of the fauna and flora of the English lakes and of those chemical and physical factors which influence abundance and distribution. It includes much of the information in the early chapters of this book. Another book (1963) covers the ground to which Chapter 8 is devoted. Hynes (1970) deals comprehensively with studies of running water all over the world. His earlier work (1960) is an authoritative review of the effect of pollution on fauna and flora. The New Naturalist series has made important contributions to the study of fresh water with its books on *The Salmon* (Jones, 1959), *The Trout* (Frost and Brown, 1967), *Dragonflies*, *British Mammals*, and *The British Amphibians and Reptiles*. Other books are mentioned in the appropriate place in the text. In the list below we have quoted only those original references which are not to be found in the books mentioned, generally because they are too recent.

To complete the picture mention must be made of *A Treatise on Limnology* (New York: Wiley) by Professor G. Evelyn Hutchinson, who is likely to be the last person to undertake the heroic task of covering the whole field. Two volumes have appeared so far and the third and last is in preparation. Ruttner's popular general work has now been translated into English (Frey and Fry 1963, *Fundamentals of Limnology* 3rd ed., Toronto U.P.).

ALABASTER, J. S. (1965). The effect on coarse fish of heated effluents from power stations. *Proc. 2nd British Coarse Fish Conference*, 109–110.

ALEXANDER, W. B., SOUTHGATE, B. A., & BASSINDALE, R. (1935). Survey of the River Tees II. The estuary—chemical and biological. *Tech. Pap. Wat. Pollut. Res. Lond.* 5: i–xiv and 1–171.

ALLEN, K. R. (1945). The trout population of the Horokiwi River: an investigation of the fundamentals of propagation, growth, and survival. *Ann. Rep. Mar. Dep. New Zealand, 1945:* 1–8.

ALLEN, K. R. (1952). A New Zealand trout stream, some facts and figures. *New Zealand Marine Dept. Fish. Bull., 10A,* 70 pp.

ALM, G. (1946). Reasons for the occurrence of stunted fish populations, with special regard to the perch. *Lantbruksstyrelsen, 25:* 1–146.

BALFOUR-BROWNE, F. (1940). *British water beetles.* London, Ray Soc., i–xx and 1–375.

BEADLE, L. C. (1939). Regulation of the haemolymph in the saline water mosquito larva *Aedes detritus* Edw. *J. Exp. Biol. 16:* 346–62.

BERG, K. (1938). Studies on the bottom animals of Esrom Lake. *K. Danske Vidensk. Selsk. Skr. 8:* 1–255.

BERTIN, L. (1956). *Eels. A biological study.* London, Cleaver-Hume. viii + 192.

BOYCOTT, A. E. (1936). The habitats of fresh-water Mollusca in Britain. *J. Anim. Ecol. 5:* 116–86.

BRUUN, A. F. (1963). The breeding of the North Atlantic freshwater eels. *Advanc. mar. Biol. 1:* 137–69.

BUTCHER, R. W. (1933). Studies on the ecology of rivers. 1. On the distribution of macrophytic vegetation in the rivers of Britain. *J. Ecol. 21:* 58–91.

BUTCHER, R. W. (1946). The biological detection of pollution. Paper read to The Institute of Sewage Purification, 118 Victoria Street, Westminster, London, S.W.1., 3–8.

BUTCHER, R. W., LONGWELL, J., & PENTELOW, F. T. K. (1937). Survey of the River Tees. III. The non-tidal reaches—chemical and biological. *Tech. Pap. Wat. Pollut. Res. Lond. 6:* i–xiii and 1–189.

BUTCHER, R. W., PENTELOW, F. T. K., & WOODLEY, J. W. A. (1931). An investigation of the River Lark and the effect of beet sugar pollution. *Fish. Invest. 3:* No. 3, 1–112.

CARPENTER, K. E. (1926). The lead mine as an active agent in river pollution. *Ann. Appl. Biol. 13:* 395–401.

CARPENTER, K. E. (1927). The lethal action of soluble metallic salts on fishes. *J. Exp. Biol. 4:* 378–90.

CARPENTER, K. E. (1928). *Life in Inland Waters.* London, Sidgwick & Jackson. i–xviii and 1–267.

DAVIES, R. W. (1969). Predation as a factor in the ecology of triclads in a small weedy pond. *J. Anim. Ecol.* 38, 577–84.

DIVER, C. (1933). The physiography of South Haven Peninsula, Studland Heath, Dorset. *Geogr. J. 81:* 404–27.

ELLIS, E. A. (1965). *The Broads.* New Naturalist No. 46. London, Collins. xii + 401.

FITTER, R. S. R. (1945). *London's Natural History.* New Naturalist No. 3. London, Collins. i–xii and 1–282.

FOREL, F.-A. (1892). *Le Léman. Monographie limnologique. 1,* 1892, 1–543. *2,* 1895, 1–651. *3,* 1904, 1–715. Lausanne, Rouge.

FORT, R. S., & BRAYSHAW, J. D. (1961). *Fishery Management.* London, Faber & Faber. 398.

FRANKLAND, E., & MORTON, J. C. (1874). *Domestic Water Supply of Great Britain.* Sixth Report of the Commissioners appointed in 1868 to inquire into the best means of preventing the pollution of rivers. Rivers Pollution Commission (1868). London, H.M.S.O. i–xii and 1–525.

FROST, W. E., & BROWN, M. E. (1967). *The Trout.* New Naturalist Special Volume 21. London, Collins. 286. (Fontana, 1970.)

GOLDMAN, C. R. (1965). Micronutrient limiting factors and their detection in natural phytoplankton populations. In Goldman, C. R. (ed). *Primary productivity in aquatic environments. Mem. Ist. Ital. Idrobiol,* Suppl. 18, Berkeley: Calif. U.P. 121–135.

GREGORY, M. (1967). *Angling and the Law.* London, Knight. xvi + 196.

HARTLEY, P. H. T. (1947). The coarse fishes of Britain. *Freshw. Biol. Assoc. Brit. Emp. Sci. Pub. 12:* 1–40.

HASTINGS, A. B. (1937). Biology of water supply. *British Museum (Natural History) Econ. Ser. 7a:* 1–48.

HEAPE, W. (1931). *Emigration, Migration and Nomadism.* Cambridge, Heffer. i–xii and 1–369.

HICKLING, C. F. (1962). *Fish Culture.* London, Faber & Faber. 295.

HICKLING, C. F. (1968). *The Farming of Fish.* London, Pergamon. viii + 88.

HOLDEN, W. S. (1970). *Water Treatment and Examination.* (Thresh, Beale and Suckling.) London, Churchill. viii + 513.

HYNES, H. B. N. (1960). *The Biology of Polluted Waters.* Liverpool U.P. xiv + 202.

HYNES, H. B. N. (1970). *The Ecology of Running Waters.* Liverpool U.P. xxiv + 555.

ILLIES, J. (ed) (1967). *Limnofauna Europaea.* Stuttgart, Fischer. xvi + 474.

JACKSON, D. F. (ed) (1964). *Algae and Man.* New York, Plenum. x + 434.

JONASSON, P. M., & MATHIESEN, H. (1959). Measurements of primary production in two Danish eutrophic lakes, Esrom So and Fureso. *Oikos,* 10, 137–167.

JONES, J. R. E. (1958). A further study of the zinc-polluted River Ystwyth. *J. Anim. Ecol.* 27, 1–14.

JONES, J. R. E. (1964). *Fish and River Pollution.* London, Butterworths. viii + 203.

JONES, J. W. (1959). *The Salmon.* New Naturalist Monograph No. 16. London, Collins. xvi + 192.

KEILIN, D. (1927). Fauna of a horse-chestnut tree (*Aesculus hippocastanum*). Dipterus larvae and their parasites. *Parasitology 19:* 368–74.

KOFOID, C. A. (1910). The biological stations of Europe. *U.S. Bur. Educ. Bull. 440:* i–xiii and 1–360.

LAURIE, R. D., & JONES, J. R. E. (1938). The faunistic recovery of a lead-polluted river in north Cardiganshire, Wales. *J. Anim. Ecol. 7:* 272–89.

MACAN, T. T. (1959). *A Guide to Freshwater Invertebrate Animals.* London, Longmans. x + 118.

MACAN, T. T. (1963). *Freshwater Ecology.* London, Longmans. x + 338.

MACAN, T. T. (1970). *Biological Studies of the English Lakes.* London, Longmans. xvi + 260.

MELLANBY, H. (1938). *Animal Life in Fresh Water.* London, Methuen. i–viii and 1–296.

MENZIES, W. J. M. (1925). *The Salmon. Its Life Story.* Edinburgh & London, Blackwood. i–xii and 1–212.

MINSHALL, G. W., & KUEHNE, R. A. (1969). An ecological study of invertebrates of the Duddon, an English mountain stream. *Arch. Hydrobiol.* 66, 169–191.

MOON, H. P., & GREEN, F. H. W. (1940). Water meadows in southern England. The land of Britain. *Report of the Land Utilisation Survey of Britain, 89:* 373–90.

MORTIMER, C. H., & HICKLING, C. F. (1954). *Fertilizers in Fishponds.* Colonial Off. Fish. Publ. No. 5. iv + 155.

MURRAY, J., & PULLAR, L. (1910). *Bathymetrical Survey of the Scottish Freshwater Lochs.* 6 vols. Edinburgh, Challenger Office.

NETBOY, A. (1968). *The Atlantic Salmon. A Vanishing Species?* London, Faber & Faber. 457.

NICOL, E. A. T. (1935). The ecology of a salt-marsh. *J. Mar. Biol. Ass. U.K. 20:* 203–61.

NORRIS, J. D. (1967). A campaign against feral coypus (*Myocastor coypus* Molina) in Great Britain. *J. appl. Ecol. 4, 1:* 191–99.

PEARSALL, W. H., GARDINER, A. C., & GREENSHIELDS, F. (1946). Freshwater biology and water supply in Britain. *Freshw. Biol. Assoc. Brit. Emp. Sci. Pub. 11:* 1–90.

PENNINGTON, W. (1941). The control of the numbers of freshwater phytoplankton by small invertebrate animals. *J. Ecol. 29:* 204–11.

PENTELOW, F. T. K., BUTCHER, R. W., & GRINDLEY, J. (1938). An investigation of the effects of milk wastes on the Bristol Avon. *Fish Invest. 4:* 1–80.

PERCIVAL, E., & WHITEHEAD, H. (1929). A quantitative study of the fauna of some types of stream-bed. *J. Ecol. 17:* 282–314.

PERRING, F. H., & WALTERS, S. M. (1962). *Atlas of the British Flora.* London: Nelson. xxiv + 432.

PHILLIPSON, J. (1966). *Ecological Energetics.* Studies in Biology No. 1, Inst. Biol. London, Arnold. pp. 57.

PUGSLEY, A. J. (1939). *Dewponds in Fable and Fact.* London, Country Life. i–x and 1–62.

PYEFINCH, K. A. (1969). The Greenland Salmon Fishery. *The Salmon Net.* No. 5. 11–19.

ROBSON, T. O. (1968). *The Control of Aquatic Weeds.* MAFF Bull. No. 194. H.M.S.O. vi + 54.

Rollinson, W. (1967). *A History of Man in the Lake District*. London, Dent. xii + 162.

Ruttner, F. (1940). *Grundriss der Limnologie*. Berlin, 1–167. (trans. Frey, D. G. and Fry, F. E. J. (1963). *Fundamentals of Limnology*, 3rd ed. Toronto U.P.)

Savage, R. M. (1961). *The Ecology and Life History of the Common Frog*. London, Pitman. viii + 221.

Schaperclaus, W. (1933). *Textbook of Pond Culture*. US Dept. Int. Fish. Leaflet No. 311, pp. 261.

Schmidt, J. (1923). Breeding places and migrations of the eel. *Nature, London, 111:* 51–54.

Schmidt, J. (1924). The transatlantic migration of the eel-larvae. *Nature, London, 113:* 12.

Sinha, V. R. P., & Jones, J. W. (1967). The Atlantic Eel problem. *Proc. 3rd British Coarse Fish Conference*, 70–73.

Smith, E. P. (1939). On the introduction and distribution of *Rana esculenta* in East Kent. *J. Anim. Ecol. 8:* 168–70.

Smith, M. (1951). *The British Amphibians and Reptiles*. New Naturalist No. 20. London, Collins. xiv + 318.

Southgate, B. A. (1965). Progress in the control of river pollution 1948–65. *London Conference of the Salmon and Trout Association*, 1965. pp. 9.

Southgate, B. A. (1969). *Water: Pollution and Conservation*. Harrow, Middx. Thunderbird Enterprises. pp. 159.

Stamp, L. D. (1946). *Britain's Structure and Scenery*. New Naturalist No. 4. London, Collins (Fontana, 1960). i–xvi and 1–255.

Stuart, T. A. (1941). Chironomid larvae of the Millport shore pools. *Trans. Roy. Soc. Edinb. 60:* 475–502.

Tansley, A. G. (1939). *The British Islands and their Vegetation*. Cambridge U.P. i–xxxviii and 1–930.

Teal, J. M. (1957). Community metabolism in a temperate cold spring. *Ecol. Monogr.* 27, 283–302.

Thienemann, A. (1950). Verbreitungsgeschichte der Süsswassertierwelt Europas. *Die Binnengewässer* B 18, xvi + 809, Stuttgart, Schweizerbart.

Thompson, D. H. (1941). *The Fish Production of Inland Streams and Lakes*. A Symposium on Hydrobiology. 206–17, Univ. Wisconsin Press.

Tucker, D. W. (1959). A new solution to the Atlantic eel problem. *Nature, Lond.,* 183, 495–501.

Turing, H. D. (1947). *Pollution 1:* 1–48; *2:* 1–40. London, British Field Sports Society.

Ullyott, P. (1939). Die täglichen Wanderungen der planktonischen Süsswasser-Crustaceen. *Int. Rev. Hydrobiol. 38:* 262–84.

Vesey-Fitzgerald, B. (1946). *British Game*. New Naturalist No. 2. London, Collins. i–xv and 1–240.

WALTON, G. A. (1943). The water bugs (Rhynchota-Hemiptera) of North Somerset. *Trans. Soc. Brit. Ent.* 8: 231–90.

WARWICK, T. (1940). A contribution to the ecology of the musk-rat (*Ondatra zibethica*) in the British Isles. *Proc. zool. Soc. Lond. A.* 110, 165–201.

WASMUND, E. (1935). Die Bildung von anabituminösem Leichenwachs unter Wasser. *Erdölmuttersubotanz Schrift. Gebiet. Brennstoff-Geologie.* 10, 1–70.

WELCH, P. S. (1935). *Limnology.* London, McGraw-Hill. i–xiv and 1–471.

WESENBERG-LUND, C. (1908). Plankton investigations of the Danish Lakes. *Danish Freshw. Biol. Lab. Op. 5.* Copenhagen, Gyldendalske, i–xii and 1–389.

WESENBERG-LUND, C. (1943). *Biologie der Süsswasserinsekten.* Copenhagen, Gyldendalske, 1–682.

WHEELER, A. (1969). Fish-life and pollution in the lower Thames: A review and preliminary report. *Biol. Conserv. 2*, 25–30.

WHIPPLE, G. C. (1927). *The Microscopy of Drinking Water.* London, Chapman and Hall, 4th ed., i–xx and 1–586.

Index